Sociology and development

The impasse and beyond

Ray Kiely
University of East London

UCL
PRESS

First published in 1995 by UCL Press

UCL Press Limited
University College London
Gower Street
London WC1E 6BT

The name of University College London (UCL) is a registered
trade mark used by UCL Press with the consent of the owner.

British Library Cataloguing-in-Publication Data
A catalogue record for this book is available from the British Library.

Library of Congress Cataloging-in-Publication Data are available

ISBN: 1-85728-195-0 HB
      1-85728-196-9 PB

Typeset in Classical Garamond.
Printed and bound by
Biddles Ltd., Guildford and King's Lynn, England.

Contents

Acknowledgements

Some of the arguments in Chapter 5 were first developed in my article of 1994, "Development theory and industrialisation: beyond the impasse", and the themes in Chapter 7 were first developed in my article of 1995 entitled "Third Worldist relativism: a new form of imperialism".

Thanks to the Department of Cultural Studies at the University of East London for its support and in particular for allowing me relief from full teaching requirements so that I could complete this work. In particular, thanks to Bill Schwarz for reading the complete draft. Also thanks to my most important critics, the students on the Perspectives on Development, Industrialisation and Development, and State Policy and Development courses at UEL from 1992 to 1994. More general thanks to my parents and the rest of my family, especially Jenny, Lynn and Shani (and the kids), for their support. Thanks also to Phil Mizen, Jane McAllister, James Nottidge, Chris Ashplant, Sanjiv Sachdev, Paul Butler, Seb Berry, Martin and Gerry Nicholson (and little Michael), Fiona Lehane, Kim Labbett, Andrew Burroughs, Andrew Branch and Kevin Taggart for their support and lack of sobriety.

1

Introduction:
The impasse summarized

The sociology of development is in crisis. The reasons are controversial and a matter of academic debate, but many writers agree that the discipline has reached something of an impasse (see, for example, Booth 1985, Mouzelis 1988, Sklair 1988, van der Geest & Buttel 1988, Corbridge 1986, 1989, 1990, Spybey 1992). Since its establishment as a separate discipline within the social sciences in the postwar world, development sociology has undergone a number of theoretical transformations, each of which attempted to overcome the inadequacies of the previous "paradigm". However, the discipline has now reached a point where theoretical and conceptual innovation has largely "dried up" (but see Schuurman 1993, Booth 1994), which has led to an impasse on at least two fronts: first, an inability to move beyond the weaknesses of old paradigms, such as Marxism and dependency theory; secondly, and ultimately inseparable from the first point, an inability to come to terms with changes in the global order in the 1970s and 1980s (see especially Corbridge 1990: 623).

This book is concerned with explaining the reasons for the impasse, and suggesting ways of moving beyond it. Its focus is on Marxist and related theories of development and "underdevelopment" because these have dominated the field for the past 20 or so years. However, I do not ignore other theories, including "anti-sociological" theories of development such as neoliberal theory. Indeed, I argue that development *sociology*'s narrow concern with the "correct method" has left it isolated from important debates, such as the "counterrevolution" in development economics (Toye 1987). Nevertheless, the core of the book concerns itself with Marxist theories. However, the main contention of the book is that the impasse is not confined to Marxist-oriented writings on development, and indeed that the impasse can be identified as a misguided attempt to construct ahistorical and technical models of development that actually *fetishize*, in Marx's sense, social reality.

Before moving on to the main body of the work, I want to present a summary history of theories in the sociology of development since 1945 and identify some of the problems associated with each of them. This summary will illustrate the reasons for the impasse, which are then briefly listed. I complete this introductory chapter by outlining the structure of the rest of the book.

Sociology and development since 1945

In post-war analyses of what came to be known as the "Third World", modernization theory was established as the orthodoxy from the 1950s to the late 1960s (Rostow 1960, Hoselitz 1960, McClelland 1961, Apter 1965). This school of thought saw development as a succession of stages through which all "societies" (defined as the nation-state) must pass on their way to "modernity". The "advanced", modern Western world was seen as the end-point, the "norm" to which all other societies aspired. This form of functionalist sociology was criticized for its analysis of the reality of the Western world (Mann 1987), but also for its failure to grasp power relations within the international economy (Frank 1966). Such a critique formed the basis of an alternative model, known as underdevelopment theory, which argued that the "advanced" capitalist world exploited the "backward" Third World via a process of surplus transfer from periphery or satellite to the core or metropole (Baran 1957, Frank 1969a,b, Rodney 1972).

Although this alternative model, known as underdevelopment theory, was never entirely convincing, it did lead to renewed analysis of the inequities of the international economic order. Structuralist theorists such as Prebisch (1959) had criticized the terms on which Third World primary producers traded with the First World, and had argued (ironically, along with modernization theory) that industrialization in the periphery was essential to alleviate this problem. However, Frank and like-minded writers (Amin 1976, Wallerstein 1974) saw capitalist industrialization as part of the problem, because it sustained links with the world economy and so facilitated a continued process of surplus transfer from the Third to First World (Frank 1983).

The contentions of underdevelopment theory were in turn challenged by the rapid industrialization of some peripheral nations in the 1970s and 1980s (Warren 1973) and by a more orthodox school of Marxist theory, which focused on the mode of production (Laclau 1971). This challenge met with a strong defence by writers sympathetic to some of Frank's ideas but not to the rigidity of his notion of development as a zero-sum game (Cardoso & Faletto 1979). This school of thought, known as dependency theory, argued that development in the periphery was taking place, but it was in some sense "dependent", and was led by the core nations.

The relationship between Marxism and these issues was ambiguous. Underdevelopment theory was clearly influenced by Marxist ideas, but was subject to strong methodological and empirical criticisms (Laclau 1971, Warren 1973, Brenner 1977). Frank was criticized for assuming rather than explaining the metropole/satellite division of the world (Dore & Weeks 1979), for his focus on trade rather than relations of production, and for his failure to explain the rapid industrialization of parts of the Third World. The response to these criticisms varied. One response was to return to the more optimistic ideas that Marx expressed about the capacity of colonialism to modernize traditional India, which led to Warren's (1980) thesis that imperialism was the "pioneer of capitalism". Others rejected this evolutionary account, and employed an approach that suggested that the effects of capitalism in the Third World were contradictory. This led to the development of the "modes of production" school, which argued that the modern capitalist mode articulated with traditional non-capitalist modes, and in doing so gained certain benefits such as cheap labour or raw materials (see Foster Carter 1978, Wolpe 1980). This concept of articulation was often combined with a theory of the "internationalisation of capital" (Cypher 1979, Browett 1985), which examined the accumulation of capital on a global scale and the forms that this took in particular regions of the world. In some areas it was said to have led to just such an articulation, while in others it promoted rapid Third World industrialization and a "new international division of labour" (Frobel et al. 1980).

These theories, especially the theory of modes of production, clearly intended to transcend the tradition–modern dichotomy and emphasize uneven development in the Third World. Modes of production theory was established as the orthodoxy in development theory for a short while in the 1970s and 1980s (see Hindess & Hirst 1975, Taylor 1979, Wolpe 1980), and it is still an influential thesis (see for instance Peet 1991). However, it has been subject to strong criticism since then (Booth 1985, Ruccio & Simon 1986, Corbridge 1986, Kiely 1992a). The main contention of the critics is that articulation theory attempts to explain both the breakdown and the preservation of non-capitalist modes of production in terms of a "logic of capital". This criticism is now accepted by many academics in the field, but there is a paucity of potential successors.

The impasse can therefore be located in the failure to find an adequate replacement for this and related theories. But there is a second justification for concentrating on theories of structuralist Marxism. The methodology that it employs exemplifies the common problems of most theories of the sociology of development since the 1950s. In this sense, it can be used as a kind of heuristic device to examine the common methodological problems of most theories of development and underdevelopment (Kiely 1992a: 5–7). I argue in the chapters that follow that this common methodological weakness is rooted in an overemphasis on structure at the expense of agency, and on the centre at the expense of the periphery.

The impasse

The impasse exists at four interconnected levels. These are (i) evolutionism; (ii) functionalism; (iii) dogmatism; (iv) the inability of theory to explain the changing reality of the world system.

Evolutionism

Evolutionary theories of what is now called "development" are as old as social science itself. Modernization theory saw Western development as the "norm", the model for all other countries to follow. Each nation-state (this category was regarded as largely unproblematic) passed through similar stages of development and ended up "converging" with the Western model. The problem with this thesis, at least in terms of its evolutionism, is that Third World societies are analyzed not on the basis of existing reality, but rather on the basis of a future Western norm. In so far as the present is analyzed in this theory, it is in terms of how far it deviates from a "promised" market-led liberal democracy. "Orthodox" Marxism replicated these errors, although in this theory the capitalist model was seen as a goal penultimate to the communist utopia (see Chapter 2).

Underdevelopment and dependency theory undoubtedly provided a strong challenge to the evolutionism of both modernization theory and orthodox Marxism. Most importantly, dependency theory undermined the latter's conceptualization of the Third World as composed of traditional nation-states awaiting Western liberation. Instead, Third World nation-states were conceptualized in terms of their incorporation into the *modern* international division of labour. It was in this way that underdevelopment and dependency theory challenged the "tradition–modern" dichotomy that stood at the heart of modernization theory. However, this was at the expense of an implicit theorization of Western development as the "norm" and Third World underdevelopment as a deviation from that norm. Third World nations could achieve "modernity" so long as Western obstacles could be eliminated (Phillips 1977, Spybey 1992: ch.2). The tradition–modern dichotomy of modernization theory was therefore replaced with a new evolutionary model in underdevelopment and dependency theory (see Chapter 3 for more details).

Modes of production theory attempted to avoid evolutionary models and, though its success in doing so was mixed, it was far more convincing than the "modernization–dependency" debate. However, it was at the cost of intensifying the second characteristic of the impasse – functionalism.

Functionalism

The problem of confusing cause and effect was present in both modernization and dependency theories. Modernization theory's approbation of institutions such as transnational companies was largely predicated on the basis of their functionality in the process of "modernization". Orthodox Marxism has long taken a view of history based on the primacy of the productive forces, and social relations (such as class) were conceived on the basis of how they contributed toward the process of the development of the productive forces (see Cohen 1978). Underdevelopment and world-systems theory, its close relative, conceived of class and state formation as a consequence of the needs of the world capitalist economy. Changes in these structures were therefore "explained" in terms of the needs of the system. The theory of articulation of modes of production intensified this problem. The preservation of non-capitalist modes of production was assumed to be the product of the needs, or reproductive requirements, of the dominant capitalist mode of production. If a non-capitalist mode was destroyed, then this too was a product of the reproductive requirements of capitalism.

The problem with all these approaches is that they assume, rather than explain, that changes take place because of the logic of the "system". This reflects the process of concept formation in these theories, which are conceived without due attention to changing events in the "real world". Such a process of "theory building" leads on to the third characteristic of the impasse.

Dogmatism

Because data are completely separated from, and secondary to, theory, each particular theory tends to be sheltered from empirical material that challenges some of the theory's key features. In practice each theory has been forced to react to challenges from scholars working within other paradigms, but there is still a tendency to dismiss empirical work that fails to utilize "the correct method" (see Hindess & Hirst 1975: ch.1, for the most notorious example). One result of this theoretical dogmatism has been to insulate sociologists from important changes in development studies, such as the rise of the New Right. Other changes, too, have been neglected as a result of this dogmatic approach.

Changes in the global order

Although this characteristic has been less emphasized by writers who have identified an impasse, the connection between this weakness and the previous ones cited is very direct. The dogmatism of so much radical development theory has hindered discussion of changing characteristics in the global order. Three developments are especially appropriate. The neoliberal "counterrevolution" and, related to this, the collapse of "state socialism" represent an important challenge to radical development sociologists. The industrialization of parts of the Third World undermines the stagnationist ideas of Frank, which were so influential on the left. Finally, the nature of Third World nationalism, which was once given so much support by the left, also represents a new challenge to received ideas about "North–South" relations. Questions can be asked about each of these developments. For example: Why has neoliberalism enjoyed such a powerful revival? Does the rise of newly industrializing countries mean that neoliberal theory was actually correct? What does the close association between nationalism and dictatorship in the Third World tell us about nationalism, "internal relations" in the periphery, and the mechanisms of North–South relations or "imperialism"?

The structure

The rest of the book will address the issues summarized above in more detail. I am concerned with two principal problems:

(i) What has caused the impasse?
(ii) In what ways can we move beyond it?

In answering these questions, the book is divided into two parts, plus a conclusion. Part I, which covers Chapters 2–4, examines the "theoretical impasse". These three chapters explore the reasons for the impasse, and I argue that these can be explained by examining the methodology of development theory. Although not offering an alternative "grand theory" of development and underdevelopment, I suggest that there is still much to be gained from employing a flexible Marxist method, which is outlined in the chapters that follow. In suggesting the continued utility of an "unorthodox" Marxist approach, I also register my reservations about certain would-be alternatives, such as "post-Marxism". The second part, which covers Chapters 5–7, examines the impasse in the context of the "new world order". My discussion in Chapters 2–4 is used as the basis to examine standard approaches to the issues of Third World industrialization (where my flexible Marxist approach is used), neoliberalism and the collapse of state socialism, and Third World nationalism. Each particular approach is criticized, and some would-be alternatives are examined. I again register my difficulties with these approaches, and suggest some alternatives.

Part I

The theoretical impasse

2

Marx and development

Is there anything more to be said about Marxism and its relationship to the processes of history and development? Such a question will bring a smile to those development sociologists who suffered the seemingly endless "mode of production" debates in the 1970s, and the validity of this question has become all the more pronounced with the "end of socialism" and, indeed, the "end of History" in the 1990s. Even those who sympathize with at least some of the central concepts of Marxist theory have called for a break from Marx (see especially Booth 1985, 1993) so that development studies can transcend the impasse, a view that, in some respects, I share. Why, then, another book in defence of some of the major themes of Marxist thought?

Given the banality of much scholarship and activism that passes for Marxism, this is an understandable question. On the other hand, it does seem strange that scholars are so fond of asking this question when there is no disdain for books about Weber, Simmel or Durkheim. This point does not answer the question of course, but it does suggest at least one fruitful reply; namely, that, just as there is more than one interpretation of Weber, so too is there more than "one Marx". This is all the more important in the case of Marx and Marxism because much of the world in the twentieth century has been ruled by regimes claiming allegiance to this particular doctrine (on these points, see Sayer 1987).

In development studies, this point is important because the debate around the impasse, which includes the collapse of socialism, is largely a debate about the key concepts of Marxist theory. It is for this reason that this book is primarily about Marxism. This does not mean, however, that east Asian industrialization in the 1960s and

11

1970s or the current environmental crisis can be explained by simple reference to *Capital* or the *Grundrisse*. Nor does it mean that the transition to capitalist relations of production in eighteenth-century England, which Marx described in volume 1 of *Capital*, should be replicated in twentieth-century Africa. This would be to confuse *method* with dogma. Unfortunately it has too often been the case that theoretical construction in the sociology of development has concentrated on the latter at the expense of the former (which, it should be stressed, is not to condemn the work of at times quite brilliant theoretically informed empirical studies). This is the major reason for the impasse identified in Chapter 1.

What a Marxist method *can* do, however, is to lay stress on the historical and social nature of "development" and how the development process is a product of particular historical and social struggles. The purpose of this book is to try to suggest ways of moving beyond the impasse by constructively employing a Marxist *method* (as opposed to a Marxist *dogma*), by both suggesting reasons for the impasse and, through empirical analysis and theoretical interpretation, suggesting ways of transcending it. In order to do this, I set out in this chapter to contrast at least some of the central differences between Marxism as dogma and Marxism as method. In order to facilitate this task, the chapter is divided into two main sections. The first outlines the "orthodox" (dogmatic) approach to Marxism, which is based on a unilinear account of history, and an account of capitalism that focuses on its progressive side *vis-à-vis* non-capitalist modes of production. In the second section, this approach is criticized and an alternative framework is suggested. Most importantly, in this section I suggest that the orthodox approach actually "fetishizes" (in Marx's sense) reality, and, as a precursor to my discussion in Chapters 3 and 4, I argue that the impasse can usefully be explained in terms of Marx's understanding of "the fetishism of commodities".

Orthodox Marxism

There are a number of versions of what I have chosen to call orthodox Marxism, and there are important differences within this broad body of thought. For example, the work of Bill Warren (1973, 1980) emerged partly as a critique of the orthodox Marxism of the Third

Communist International under Lenin and then Stalin, but I will broadly identify both traditions as belonging to the orthodox school of Marxist thought. Similarly, the work of John Sender and Sheila Smith (1985, 1990) starts out as a critique of orthodox Marxism, but I will argue that their approach is formulated within a similarly restricted framework. Even the Althusserian-influenced development sociology examined in Chapter 4 can, in some respects, be identified as a constituent part of this orthodox tradition.

So, having identified the wide parameters of "orthodox Marxism", I will now spell out the characteristics that unite this approach. Three features can be identified: first, a unilinear account of history; secondly, an optimistic assessment of the "modernizing influence" of capitalism and colonialism; and thirdly, an assertion that the development of the productive forces is the main agency of history. Different parts of the orthodox tradition place different emphases on each of these factors, and some might even reject at least one of them (the Third International, for instance, was no apologist for colonialism). Of the three factors identified, all orthodox Marxists share a commitment to the third proposition.

The relationship between orthodox Marxism and development sociology (and how this relates to the impasse) is spelt out more clearly in Chapters 3 and 4, and so for the rest of this chapter I want to examine the origins of this orthodoxy in Marxist thought and provide the basis of an alternative approach from within the same tradition. This section examines the main contentions of orthodox Marxism, and the next section criticizes this approach from a Marxist perspective. The rest of this section is divided according to the three characteristics cited above.

A unilinear account of history

Orthodox Marxists all tend towards a "stagist" version of history, whereby all societies (which are usually assumed to be nation-states) pass through similar stages of development. Parallels are often made with the natural world, and just as the life-span of a plant goes through clearly identifiable stages of development, so too do human societies. The Marxist historical study of society, known as historical materialism, is thus nothing more than the extension of the principles of the natural world ("dialectical materialism") to the social world.

Therefore, according to this account, communism is the culmination of an evolutionary logic that is independent of human will, which arises out of the progressive nature of the capitalist mode of production. Thus, according to Marx (1977: 390):

> In broad outlines Asiatic, ancient, feudal, and modern bourgeois modes of production can be designated as progressive epochs in the economic formation of society. The bourgeois relations of production are the last antagonistic form of the social process of production.

This stagist account of history was consistently repeated in "Marxist–Leninist" texts in the former Soviet Union. For example, in the 1963 text "Fundamentals of Marxism-Leninism" (cited in Larrain 1986: 55) the writers state that:

> All peoples travel what is basically the same path . . . The development of society proceeds through the consecutive replacement, according to definite laws, of one socio-economic formation by another.

The implications of these approaches for development studies should be clear. "Backward" or peripheral societies are at a pre-capitalist stage of development. The task for Marxists is to support the promotion of capitalism in the periphery, which will in the long term promote the material conditions for a transition to communism. It is for this reason that some versions of orthodox Marxism take a very partial view of both colonialism and the capitalist penetration of the periphery.

Capitalism and colonialism as agents of modernization

Marx is perhaps best known as the nineteenth century's greatest critic of capitalism. However, as many development writers point out, he was also one of its greatest admirers. As should be clear from the above statement, Marx saw capitalism as far more progressive, and revolutionary, than any previous mode of production that had existed in history. Once again, Marx (1967: 37) is quite clear on this point:

Only the capitalist production of commodities revolutionizes the entire economic structure of society in a manner eclipsing all previous epochs.

It is a short step from this notion that capitalism represents "progress" in history to the view that capitalist penetration of pre-capitalist modes of production should be supported. It was on this basis that Marx and Engels supported the colonization and annexation of the non-capitalist world by the capitalist powers. For example, Marx in his early writings (Marx & Engels 1974: 40) supported British colonialism in India, arguing that "English interference . . . produced the greatest, and to speak the truth, the only *social* revolution ever heard of in Asia". The modernizing influence of Western capitalism was contrasted with "backward" India, which "has no history at all . . . What we call its history, is but the history of the successive intruders who founded their empires on the passive basis of that unresisting and unchanging society" (ibid.: 81). This led Marx to argue that "England has to fulfil a double mission in India: one destructive, the other regenerating – the annihilation of old Asiatic society, and the laying of the material foundations of Western society in Asia" (ibid.: 82). Engels supported the United States' annexation of parts of Mexico for similar reasons (see Larrain 1986: 86), and argued that the failure of democratic revolutions in Europe in 1848 could be attributed to the counterrevolutionary role of "non-historic nations" (see Lowy 1977: 138–40).

These views were broadly shared by the major theorists of the Second Socialist International (see Lowy 1981: ch.2) but were later completely reversed by the Communists in the Third International, who argued that imperialism held back development in the colonies and semi-colonies (see Warren 1980: ch.3).

It should now be clear that Marx at times, and many of his followers, argued that capitalism was a progressive stage of history through which all societies must pass. The question that must now be addressed is why capitalism is considered a progressive mode compared with all previous epochs of production. In answering this question we arrive at the "unifying theme" in orthodox Marxism, which is its focus on the development of the productive forces.

History as the development of the productive forces

Orthodox Marxism has a number of competing strands, and there have been a wide variety of debates between these strands on questions of Marxist theory and socialist strategy. For instance, the Bolsheviks led a socialist revolution in Russia in 1917, whereas the more openly "stagist" Mensheviks argued that Russia first required a capitalist transformation before it could become "ripe" for the creation of a socialist society. These were obviously key differences in socialist strategy, but I will try to show that the Bolsheviks and Mensheviks can both be considered part of the orthodox tradition of Marxist thought. This orthodoxy stems from a particular conception of the development of the productive forces throughout history.

In examining this theory of the productive forces, I make a distinction between two related contentions. The first argument is that there is a tendency for the productive forces to develop throughout history. The second contention is that capitalism is progressive because it develops the productive forces more rapidly than any other mode of production. After examining these approaches I briefly examine the contention of the Third International that imperialism holds back the development of the productive forces, and argue that this is simply a mirror image of Marx's worst writings on colonialism in India and elsewhere (see above).

The argument that there is a tendency for the productive forces to develop in history (for the best exposition, see Cohen 1978) rests on a particular conception of the relationship between the forces of production and the relations of production. History is regarded as a process where the productive forces (which can be broadly defined as means of production, such as technology and raw materials, plus human labour – see Cohen 1978: 32) have a tendency to develop, and there arise corresponding relations of production, which function to promote the continued development of the productive forces. Once these relations become "fetters" on the forces of production, then they are replaced by a new set of relations of production that restore the tendency of the productive forces to develop in history. As Marx (1977: 390) so (in)famously argued in 1859:

> No social order ever perishes before all the productive forces for which there is room in it have developed: and new, higher rela-

tions of production never appear before the material conditions of their existence have matured in the womb of the old society itself.

Marx had argued along similar lines in *The poverty of philosophy*, where he argued that "[t]he handmill gives you society with the feudal lord; the steam-mill, society with the industrial capitalist" (1976a: 102). The orthodox tradition takes this correlation between technology and society and argues that the former actually causes the latter; that is, the development of society is ultimately determined by the level of development of the productive forces. Marxism is therefore regarded as a "technological determinist" account of history (see Cohen 1978: 30–1).

This account of history was established as the orthodoxy in Marxist theory during the years of the Second International (1889–1914). Its leading theorists, such as Karl Kautsky and George Plekhanov, on the whole shared a theory of Marxism as a technological determinism. For instance, Plekhanov (1976: 33) argued that "[t]he organization of any given society is determined by the state of its productive forces. As this state changes, the social organisation is bound to change too." Kautsky was even more explicit when he argued that "technical progress constitutes the basis of the entire development of humankind" (cited in Larrain 1986: 46).

This technological determinism conformed to the unilinear account of history cited above, whereby all nation-states pass through similar stages of development on their way to a predestined communist future. This account in turn led to a stagist political strategy in which Marxists gave their support to the development of capitalist relations of production in "backward" countries. It was for this reason that Marxists often supported colonialism, and it was for similar reasons that the Second International foresaw only a capitalist revolution in backward countries like Russia (at least for the immediate future). This view was not only held by Plekhanov, the "father of Russian Marxism", and the Mensheviks, but was also shared (albeit with some ambiguities) by Lenin's Bolsheviks before 1917. For instance Lenin (1977: 75) wrote in his 1906 work *Two tactics of Social-Democracy in the Democratic Revolution* that:

Marxists are absolutely convinced of the bourgeois character of the Russian revolution. What does this mean? It means that the democratic reforms that become a necessity for Russia, do not in themselves imply an undermining of capitalism, the undermining of bourgeois rule; on the contrary, they will, for the first time, really clear the ground for a wide and rapid, European and not Asiatic, development of capitalism; they will, for the first time, make it possible for the bourgeoisie to rule as a class.

I return to Bolshevism and its contradictions below. For now, it should be clear that orthodox Marxism has a particular conception of history, which is based on the primacy of the productive forces. In allowing for the development of the forces of production, societies must pass through a capitalist stage of development on the way to a future classless society. I have already established that, for Marx, capitalism was a progressive mode of production compared with all previous modes in history. The question that must now be asked, then, is why this is the case. Or, to put the question another way, in what respect does capitalism provide the material basis for a future communist Utopia?

The answer to this question refers to the specific way in which capitalism develops the productive forces. According to orthodox Marxism, pre-capitalist relations of production functioned in history to develop the productive forces, but it is capitalist relations that most effectively develop the productive forces in history. It is at this point that the technological determinism of orthodox Marxism may actually break down, because there is a strong implication that it is capitalist *relations of production* that develop the *forces of production*, rather than vice versa. Nevertheless, orthodox Marxism interprets this development in a particular (and unilinear) way, so that capitalist relations are assessed on the continued orthodox basis of their functionality to developing the productive forces.

Capitalism is therefore unique in that it leads to an unprecedented development of the forces of production. This is because capitalism is based on the creation of specific relations of production, whereby the exploited class (the proletariat) is deprived of direct access to the means of production, and so is forced to work for the owners of the means of production (the capitalist class) in order to live. The details of this process are examined in the next section. For the moment, it should be stressed that these relations of production in turn give rise to the generalization of production for the market (commodity production). In other words, pre-capitalist societies were based firstly on production. Commodity production certainly existed, but it was secondary to production for direct use. In capitalist society, this is no longer the case and so goods are produced for a competitive market. This in turn has enormous implications for the development of the productive forces, because production is now geared to a market place, and this gives rise to competition between the owners of the means of production. Therefore, to stay ahead of competitors, individual capitalists are compelled to invest in new technology, increase labour productivity and search for new markets, otherwise they risk being left behind and going out of business. Writing in *The Communist manifesto*, Marx (1977: 224) was quite clear on this point:

> The bourgeoisie cannot exist without constantly revolutionizing the instruments of production, and thereby the relations of production, and with them the whole relations of society. Conservation of the old modes of production in unaltered form, was, on the contrary, the first condition of existence for all earlier industrial classes. Constant revolutionizing of production, uninterrupted disturbance of all social conditions, everlasting uncertainty and agitation distinguish the bourgeois epoch from all earlier ones.

It is for this reason that Marxists have often given their support to colonialism and imperialism. As stated above, the modernizing force of capitalism, and its tendency to develop the productive forces in an unprecedented way, was contrasted with the stagnation of pre-capitalist, stagnant, "non-historic" societies. According to this interpretation of Marxism, imperialism and colonialism may be exploitative but they are also necessary. Marx (1973: 320) was again characteristically clear about this:

> I know that the English millocracy intends to endow India with railways with the exclusive view of extracting at diminishing expenses the cotton and other raw materials for their manufacturers. But when you have once introduced machinery into the locomotion of a country which possesses iron and coals, you are unable to withhold it from its fabrication . . . The railway system will therefore become, in India, truly the forerunner of modern industry. [For a contemporary version of this argument, see Warren 1980: ch.2]

It should by now be clear that orthodox Marxism sees capitalism as a necessary stage of development in humanity's path towards a communist future. The question that must now be asked is, why? According to orthodox Marxism, capitalism is basically progressive because it provides the material conditions for a communist future. It does so in two ways. First, as should by now be clear, it develops the productive forces on a scale that is unprecedented in history. This leads to the prospect that human beings can be emancipated because they are no longer subject to the whims of nature, but are potentially in control of it. In other words, capitalism creates the possibility of abundance for all and therefore the liberation of human beings from the "necessity" of class exploitation. As Cohen (1978: 24) argues:

> So much technique and inanimate power are now available that arduous labour, and the resulting control by some men over the lives of others lose their function, and a new integration of man and nature in a new communism becomes possible.

Thus, once again, the primacy of the productive forces and the progressive nature of capitalism are stressed. However, capitalism's tendency to develop the productive forces is only one side of the story, because Marxists recognize that capitalism is a class society. What is therefore needed is the existence of a class that will overthrow capitalism and therefore *actually realize* the *potential* for liberation opened up by the development of the productive forces. It is at this point that the second key factor enters the picture, and that is the creation of a working class in capitalist society. Orthodox Marxists see this class as the key revolutionary class because it is the only class that has a vested interest in, and the power to overthrow, capitalist society. Unlike previous exploited classes in history, the proletariat is united in the process of production by modern machinery, which leads to an increase in its size, unity, strength and power (Marx 1977: 228). These factors make the workers a truly unique class in history, as capitalism, by virtue of its development of the forces of production, simplifies class antagonisms, and thus unintentionally promotes working-class resistance. Marx (1977: 229) therefore argued that:

> Of all the classes that stand face to face with the bourgeoisie today, the proletariat is a really revolutionary class. The other classes de-

cay and finally disappear in the face of Modern Industry; the proletariat is its special and essential product.

So, to return to the orthodox version of historical materialism, capitalism develops the productive forces but in the process also creates a class that will eventually overthrow class society. For the first time in history (at least since "primitive communism"), the material basis exists for a classless society because capitalism eliminates scarcity and thus the functional need for social classes. The class that fulfils this "historic mission" is the working class, which overthrows its exploiters, the bourgeoisie, and thus creates the "subjective basis" for communism.

This, in a nutshell, is the orthodox version of Marxist theory. Since the effective demise of the Second International in 1914, it has undergone some revisions, which I will briefly mention here (they are examined in more detail in Chapters 3 and 4). The most important change occurred in the context of the Russian revolution in 1917. This was a socialist revolution in a backward country led by a communist party (the Bolsheviks). Some of the old dogmatists clung to their old beliefs, and denounced the revolution as a sham. For instance, Plekhanov on his death-bed in 1918 asked his friend Leo Deutsch: "Did we not start the Marxist propaganda too soon in this backward, semi-Asiatic country?" (cited in Lowy 1981: 34). The Bolsheviks, on the other hand, had obviously broken with their previously unilinear conception of history and stagist political strategy, and in 1917 adopted a political position that was close to Trotsky's strategy of "permanent revolution" (see Liebman 1980: 182, Lowy 1981). However, this break with stagism was only partial, and the Bolsheviks remained committed to an orthodox Marxism because they continued to stress the primacy of the productive forces. This was reflected in their adoption of capitalist techniques within the workplace, and in Stalin's brutal emphasis on the need to overtake the West in terms of industrial output. More relevant here, it was also reflected in the Bolsheviks', and the Comintern's, conceptualization of the nature of imperialism. Marx's optimism concerning the transformation of India was completely overturned by the Third International, which argued that imperialism was actually an *obstacle* to the development of the productive forces in the colonial and semi-colonial countries (see Warren 1980: ch.4). At first sight, this "about-turn" might be seen as a challenge to orthodox conceptions, but it should actually be located within the same problematic, which is the almost exclusive focus on the nature of the productive forces. Imperialism is measured solely on the grounds that it either develops, or fails to develop, the productive forces. If it develops them, then it is seen as progressive; if it fails to develop them, then an alternative agency must be found to fulfil what Marx had seen as capitalism's "historic mission". In the case of the Bolsheviks and the Comintern, especially under Stalin, these alternative agencies were the state (within the Soviet Union) or the "anti-imperialist national bourgeoisie" (within the colonies and semi-colonies). So, the agency changes, but the basic focus remains the development of the productive forces.

To summarize: orthodox Marxism is based on an account of history that sees the development of the productive forces as the moving force of history, and that assesses relations of production on the basis of their functionality to developing the former. So, writings by the early Marx, the leading theorists of the Second International and Bill Warren all share the view that imperialism is the best promoter of development in the "periphery", while the Bolsheviks and the Comintern (and, I argue in Chapter 3, underdevelopment theory) argue that the state or the "national bourgeoisie" are the most effective agents. Warren's work (1980), which is a critique of the Comintern and underdevelopment theory, can be situated within exactly the same orthodox Marxist problematic, a point I expand on in the next chapter.

These arguments of course have serious implications for any discussion of the contemporary "developing world". I take up this point in later chapters, but what I can make immediately clear is that much of the impasse in radical development theory is a return to these debates and the search for an agency that will develop the productive forces. (This also applies to right-wing approaches to development, as I show later, which sees the "market" as the most effective agency for facilitating this process.) It will be clear from my discussion below that I reject such approaches to the study of development, but one point may be said in its defence, which is that the development of the productive forces is not in itself something that is to be rejected on *prima facie* grounds. Any worthwhile study of development must analyze and explain the relationship between economic growth and the wider process of development. However, what *should* be rejected is a "model" of development that focuses exclusively on the need to develop the productive forces, and that measures every other factor on the basis of its ability to undertake this task. This approach, I argue in the next section, fetishizes Marx's method and his interpretation of history.

A Marxist critique of orthodox Marxism

The approach to Marxism outlined above is now widely rejected in social science, although it still enjoys some influence in development studies (see Chapter 3). It has rightly been identified as an example of the worst kind of Euro-centric, modernist arrogance, in which the "superior" West looks at the "inferior" Rest as a backward, stagnant and incomprehensible "other" (Said 1978: 153–6, Hall 1992). More generally, and not unrelated to this point, it has also been argued that Marxism rests on a narrow account of social change, which is rooted in a technological approach to the study of history (Laclau & Mouffe 1985: 77–8). It is therefore a short step from arguing for the primacy of the productive forces to arguing that the West, where these forces are most developed, is the "civilizing agent" for the Rest.

In this section I want to reject the orthodox Marxist account outlined in the previous section. However, rather than propose a "post-Marxist" alternative (Laclau & Mouffe 1985), I will argue that, notwithstanding Marx's frequent fall into the depths of modernist conceit, orthodox Marxism is largely a caricature, and indeed a fetishization of Marx's method. This will be done by dividing this section into three subsections. First, I will examine some of Marx's writings which depart from the orthodox Marxist approach, and show how these challenge his own modernist assumptions. Secondly, I will look again at Marx's account of the transition from feudalism to capitalism in England, and assess the status of this work in Marx's thought – and in particular its applicability (or otherwise) to other societies. Thirdly, I argue that Marx's account of fetishism can itself be applied to orthodox Marxism, and, as a precursor to the rest of the book, I argue that this approach is a useful starting point in explaining the impasse in development studies.

Marx against orthodox Marxism

Marx was in many respects a modernist, but he was certainly not unaware of the conflicts created by the modern world. Moreover he did not reduce all of these to the "essential" conflict between the Western bourgeoisie and the Western proletariat. In fact, Marx's work contains the seeds of elements of the "dependency" perspective that became so influential in development theory from the late 1960s (see Chapter 3). In volume 1 of *Capital* (1976b: 925) he argued that "the veiled slavery of the wage earners in Europe needed the unqualified slavery of the New World as its pedestal". And as early as 1847 Marx had argued that there was a close link between capitalism and slavery because "[w]ithout slavery you have no cotton; without cotton you have no modern industry" (Marx 1977: 203).

Marx was also fully aware that integration into the world market was not on its own sufficient to undermine "archaic" and "backward" forms of production. Again in *Capital* (1976b: 345) he wrote that "as soon as people, whose production still moves within the lower forms of slave-labour, the corvee etc. are drawn into a world market dominated by the capitalist mode of production, whereby the sale of their products for export develops into their principal interest, the civilized horrors of over-work are grafted onto the barbaric horrors of slavery, serfdom, etc."

These comments (which pervade much of Marx's work) suggest that Marx's views on colonialism were not straightforward. Although it remains true that Marx exaggerated the stagnant character of supposedly traditional societies (on India, see Alavi 1989), it is also the case that his early apologies for colonialism were substantially altered towards the end of his life. In 1853, he had argued that the expansion of the railway system in India would pave the way for the development of industry (and the productive forces); by 1879 he was arguing that "the railways gave of course an immense impulse to the development of Foreign Commerce, but the commerce in countries which export principally raw produce increased the misery of the masses" (Marx & Engels 1974: 298). By 1881 Marx was even more explicit,when he argued that the abolition of the communal ownership of land in India "was only an act of English vandalism which pushed the indigenous people not forward but backward" (cited in Larrain 1989: 49). He also described the "bleeding process" whereby the British colonial administration extracted resources from India for the benefit of the metropolis (Marx & Engels 1974: 340). Such views were entirely consistent with Marx's earlier calls for national independence in Ireland (ibid.: 325–6). So, Marx was no simple apologist for colonialism. What these quotations suggest – although they do no more than this – is that an analysis of what is now called "modernity" cannot focus solely on ("traditional") nation-states, but must also take account of a specifically *modern* international division of labour. This alternative analysis may also suggest that a unilinear account of history, in which nation-states pass through similar stages of development, is inadequate. I return to these points below.

The significance of the transition

In the first section above I briefly outlined the reasons why capitalism has a tendency to develop the productive forces in an unprecedented fashion. I now want to explain this tendency in more detail, by examining the transition from feudalism to capitalism in England. After outlining this process, I will assess both the "orthodox" and "unorthodox" accounts of the significance of this process.

The key to understanding the decline of feudalism and the rise of capitalism is the class struggle between lord and peasant. This took the form of lords striving to increase the surplus from peasants, so that they could improve their position as rulers. On the other hand, peasants resisted this process, through either trying to enforce a reduction in rent, increasing the productivity of the land, or enlarging the land-holding without a corresponding increase in rent (Hilton 1976b: 116). It was on this basis that surplus production under feudalism increased, which in turn led to an increase in commodity production, and even in international trade. The result was an increase in the significance of the market, which further hastened the differentiation of the peasantry. This differentiation led to the slow development of a small class of capitalist farmers, which employed a steadily growing class of wage-labourers, who were displaced from the land by these same developments.

Despite peasant resistance, in England the landlords maintained control of the land. This victory for the landlord class paradoxically led to an intensification of the processes described above. From as early as the sixteenth century, small landowners began gradually to disappear, as landlords, using their political power, enclosed demesnes and vacant peasant plots (Hobsbawm 1968: 80). This process was intensified from the eighteenth century, when "the law itself now becomes the instrument by which the people's land is stolen" (Marx 1976b: 885, see also Thompson 1963: 237–43). This development had the effect of increasing the number of landless labourers, and therefore increasing the labour supply available to capitalist tenant farmers. In the long run, this meant that peasants were proletarianized, and, by one and the same act, commodity production was generalized, which gave rise to the competitive accumulation of capital.

So, for Marx (1976a: 874), the transition from feudalism to capitalism, the so-called primitive accumulation of capital, "is nothing else than the historical process of divorcing the producer from the means of production". This does not preclude colonial plunder as a factor that contributes to capital accumulation, as the quotations from the "late Marx" above make clear, but this process of "surplus extraction" on its own cannot facilitate a transformation to capitalism (see Brenner's still unsurpassed 1977). It is the emergence of the capital–labour relation that is the key to the emergence of capitalism, and it is ultimately this relation that leads to important changes in the development of the productive forces. As Brenner argues (1986: 42), "it is the capitalist property relations per se which account for the distinctive productiveness of modern economies – not any particular advance in the productive forces – and this is because capitalist property relations impose the requirement to specialize, accumulate, and innovate or go out of business."

This contrasts with non-capitalist modes where there *may be* some development of the productive forces, but there is not the same tendency to continuous revolutionizing of the means of production. Again Brenner (1986: 28) makes this clear:

> in allowing both exploiters and producers direct access to their means of reproduction, pre-capitalist property forms (as "patriarchal forms") freed both exploiters and producers from the *necessity to buy* on the market what they needed to reproduce, thus of the necessity to produce for exchange, thus of the necessity to sell competitively on the market their output, and thus of the necessity to produce at the socially necessary rate.

It should already be clear that, contrary to the claims of orthodox Marxism, the relations of production cannot be analyzed solely on the basis of their functionality to the development of the forces of production. I return to the implications of this point in a moment, but first I should comment again on how orthodox Marxism relates to this transition process. Although Marxists may accept the account of the transition outlined above (even though it undermines the thesis of the primacy of the productive forces), the implications of this transition process in England are "mapped on" to an analysis of the transition process elsewhere. So, because capitalist relations in agriculture promoted the development of the productive forces in England, Marxists should support the promotion of capitalist relations in agriculture elsewhere. Marx's account of the genesis of capitalism (1976b: chs 24–32) in English agriculture is therefore "imposed" on other countries. In other words, in this case orthodox Marxism asserts the primacy of the productive forces "through the back door". This in turn leads to a stagist theory of history (because all countries must pass through the capitalist stage) and an apology for colonialism and imperialism (because they break up pre-capitalist modes of production). For contemporary accounts along these lines, see Warren (1980) and Sender & Smith (1985).

The first section above showed that there is considerable support for these views in Marx's writings, as well as in the writings of "classical Marxists". In the Preface to the first edition of *Capital*, Marx (1976b: 91) clearly spelt out a stagist view of history when he wrote that "[t]he country that is more developed industrially only shows, to the less developed, the image of its own future". However, I have shown in this section that Marx also criticized a unilinear account of the historical process. Indeed, in his statements on the transition from feudalism to capitalism he was most explicit on this point. In his famous (but still neglected) letters to the Russian Marxist, Vera Zasulich, Marx stated that his account of the evolution of capitalism was "*expressly* restricted to *the countries of Western Europe*" (Marx 1984: 124). Marx also criticized those "Marxists" who "insist on transforming my historical sketch of the genesis of capitalism in western Europe into a historico-philosophic theory" (Marx 1982: 109–110).

These comments, along with others cited above, suggest that Marx was not consistent in his account of history and the impact of capitalism. However, a number of comments made towards the end of his life, some of which I have cited, suggest that, if there was any consistency, it was toward the view that the transition was unlikely to be replicated elsewhere (see the brilliant – and still ridiculously neglected – essays by Shanin and Wada in Shanin 1984). It was in the light of this break from stagism (or more accurately from an inconsistent vacillation between stagism and a more flexible view of history) that Marx altered his views about *class* relations, as well as the forces of production. If it was unnecessary for all countries to pass through a capitalist stage of development, then, by implication, the working class is not necessarily the only class that can lead the struggle for socialism. In the Preface to the Russian edition of *The Communist manifesto*, Marx's last writing published during his lifetime, he wrote (1977: 584) that "the present Russian common ownership of land may serve as the starting-point for a communist development". While it remains true that he attached certain conditions to this statement (namely, revolution spreading to the West), the fact remains that Marx was now arguing that a *socialist* revolution could take place in a "backward" country, and so it did not have to pass through a capitalist stage of development. Moreover – and this is my point about Marx's changing views on class – the "common owners" of the land in Russia were *peasants*, not workers. So, this statement alters the standard Marxist account of *both* the productive forces and the relations of production.

So, textual evidence suggests that Marx was clear that his account of the transition from feudalism to capitalism was intended not as a general theory of development, but as a theoretically informed historical analysis of the process in England. Textual evidence is one thing however; more important of course is whether Marx was correct to make these statements. I believe that there are strong grounds for arguing that Marx's rejection of stagism was indeed correct, and that orthodox Marxists are actually guilty of "fetishizing" Marx's account of the transition from feudalism to capitalism. What I mean by this statement should become clearer as my argument progresses but, in a nutshell, orthodox Marxism transforms Marx's account of the transition from a historical *process* into an ahistorical *model*. This is far from unique to orthodox Marxism, as I make clear in the chapters that follow.

Orthodox Marxism and fetishizing Marxist categories

Analysis of the transition process in England is of such great significance because it demonstrates that capitalism, rather than being the natural "state of things", is actually a product of social struggles rooted in English history. As Marx (1976b: 273) argues:

> Nature does not produce on the one hand owners of money or commodities, and on the other men possessing nothing but their own labour-power. This relation has no basis in natural history, *nor does it have a social basis common to all periods of human history*. It is clearly the result of a past historical development, the product of many economic revolutions, of the extinction of a whole series of older formations of social production.

Two comments can be made about Marx's argument here. The first point is that there is no good reason to expect anything else other than variety. The quotation above makes it clear that Marx was arguing that transitions are not products of a uniform process but are "bound up with certain *historically specific* patterns of the development of the contending agrarian classes and their relative strength in the different European societies: their relative levels of internal solidarity, their self-consciousness and organization, and their general political resources" (Brenner 1986: 36). This means that no particular transition can be taken as a model for another, because each transition has its own particular history of struggle. Orthodox Marxism, on the other hand, *abstracts* from these historically specific social struggles, and so presents a fixed model for all societies, irrespective of time and place. In the process, transitions are no longer seen as *social forms*, but are instead seen as natural models. Thus, to return to Cohen (1978), he argues that new relations of production will emerge on the basis of their functionality to the productive forces. But this again "fetishizes" the transition, conflating its historical emergence as a social form with its necessity to develop the forces of production. As Meiksins Wood (1984: 104) argues, "[w]hen Marx speaks of the 'historical task' of capitalism, he is not identifying the causes or explaining the processes that gave rise to capitalism; he is making a statement about the *effects* of capitalist development". The emergence of capitalism rested on the emergence of a specific class relation, which may, or may not, emerge elsewhere. This will depend on the particular balance of class forces, and this cannot be determined *a priori* but can be examined only on an empirical basis.

Hence my second, closely related point concerning Marx's statement above: since Marx's day, he has been proved right, and the transition to capitalist relations in agriculture has taken different forms in different places. Indeed, in some "advanced" capitalist countries such as the USA and France (until 1945, or even later), the capital–labour relation is the exception rather than the norm in agriculture. This observation is even more true in the so-called Third World, including in the most successful newly industrializing countries in east Asia, as I show in Chapter 5 (for a very useful survey of agrarian transitions, see Byres 1991).

It is for these reasons that orthodox Marxism actually fetishizes key categories of Marx's thought. According to Marx (1976b: 165), in capitalist society the relationship between human beings "assumes . . . the fantastic form of a relation between things". This is the basis for his analysis of "commodity fetishism", in which the products of social activity take the form of appearance of being natural phenomena. Therefore, fetishism can be defined as the conflation of the "material and formal or natural and social" (Sayer 1987: 40). This can be seen most clearly in the case of money, which plays a key *social* role in any economy where exchange exists, but actually appears as a *natural* phenomenon. As Marx states (cited in Sayer 1987: 41) "a social relation, a definite relation between individuals, appears as a metal, a stone, as a purely physical, external thing".

This "naturalization of the social" also entails a second fetishism, which is the "universalization of the historical". If one naturalizes social phenomena then one automatically strips them of any meaningful historical content. The consequence of such fetishism is that capitalist relations of production are both naturalized and dehistoricized. So, to return to the example above, the transition from feudalism to capitalism loses its *historical* and *social* significance, and is therefore seen as an *ahistorical* and *asocial* model. This in turn leads to a peculiarly narrow definition of class, which is reduced to a purely "economic" and technical concept, confined to the immediate process of production, and, as a consequence, to the assumption that capitalist rationality is eternal. It is this vulgar Marxism that has rightly been criticized by post-Marxists (Hindess & Hirst 1977, Laclau &

Mouffe 1985), but it is a version of Marxism that breaks with key concepts in Marx's thought.

Orthodox Marxism (and, I will argue in Chapter 4, structuralist Marxism) reifies key categories of Marx's thought, and in particular the category of the productive forces. To argue that there is a tendency for the productive forces to develop in history fetishizes Marx's method, with the result "that capitalism is assumed in order to explain the onset of modern economic growth, while precapitalist property relations somehow magically disappear" (Brenner 1986: 36, see also Clarke 1980, Duquette 1992). This fetishizes the capitalist mode of production because it becomes a natural, universal phenomenon, rather than a historical and social one. The alternative scenario within orthodox Marxist thought, discussed above, is to accept the importance of capitalist relations of production for the development of the productive forces but then to posit the English case as a "model" for the developing world. But this account also fetishizes the capitalist mode of production, because relations of production are "measured" solely on the basis of their utility in developing the forces of production, and in isolation from the real, concrete social struggles to be found in different places at different times (Larrain 1986: 81).

Thus, in rejecting orthodox Marxism and focusing on economic development as the outcome of *historically specific* class struggles, one is simultaneously rejecting the idea that these outcomes can simply be mapped on to other societies. Therefore, there must be a considerable amount of *indeterminacy* (Brenner 1986: 36) or uncertainty in any convincing Marxist account. Marx himself made this clear in his late writings, as I have already showed. Historical materialism is therefore best characterized as a "middle range theory", not by a theory of history "in general" (orthodox Marxism) but as a general theory of history. What this means, contrary to the interpretation of Marx's leading critics (see Foucault 1980, Popper 1986), is that Marxism does not propose "a philosophy or transcendental account of the necessary course of history but is concerned with history in the sense that it constructs the concepts necessary to render historical processes intelligible" (Larrain 1986: 98). Meiksins Wood (1984: 105) is worth quoting at length on this point, because her clear statement informs the methodology of the rest of the book:

Marxist theory can point us in the direction of class struggle as the operative principle of historical movement and provide the tools for exploring its effects, but it cannot tell us *a priori* how that struggle will work out. And indeed why should it? What Marxist theory tells us is that the productive capacities of society set the limits of the possible, and, more specifically, that the particular mode of surplus extraction is the key to the social structure. None of this makes history accidental, contingent or indeterminate. For example, if the outcome of class struggle is not pre-determined, the specific nature, conditions, and terrain of struggle, and the range of possible outcomes, certainly are historically determinate.

Capital should therefore be seen not as an *a priori* theoretical work, based on a general theory of transition, but as an account of capitalism as a social phenomenon, and how certain forms – in particular, the commodity – appear as natural but are in fact the product of *historically constructed social relations* (see Williams 1978b). One must therefore distinguish between the historical and social methodology used in *Capital* (which may be useful in understanding other societies), and the assumption that Marx's work is an account of how every society develops, irrespective of time and place. So, to take one example, "laws of motion" may exist in any particular mode of production (see Chapter 5 for one example), but these are not metaphysical laws that somehow exist beyond human agency – instead, they should be seen as historically and socially constructed laws, specific to a particular period of history.

I return to these questions throughout the text, and most specifically in Chapter 4, where I take up the question of structuralist Marxism and development theory. What should be clear by now is that Marx, at least when he was consistent with his own method, was not proposing a Euro-centric model for the rest of the world to follow. Instead, he was explaining why capitalist relations emerged in one particular part of the world, and, by implication, how this process facilitated the dominance of the rest of the world. Moreover, his work was a *critique* of existing categories of political economy, which took for granted (fetishized) that which needed explaining – that is, the emergence of generalized commodity production.

I have argued that orthodox Marxism can be criticized on the very same basis. Productive forces are "fetishized" as natural "things", while relations of production are regarded as relations of ownership between individuals and "things". In both cases the concepts are reduced to their forms of appearance in capitalist society – in other words, transhistorical definitions of relations and forces of production are superimposed on specifically historical categories, and so in the process capitalism is dehistoricized and naturalized (see Sayer 1987: ch.6, also, more generally Colletti 1972, Ollman 1976).

Conclusion: Marx, orthodox Marxism and development studies

The discussion in this chapter suggests that Marxism may remain a useful approach to the study of contemporary societies. What I have tried to show is that orthodox Marxism, in presenting a unilinear account of history or stressing the necessity of finding an agency to develop the productive forces, in many respects actually fetishizes important parts of Marx's thought (Sayer 1987). What this chapter is *not* stating is that Marxism has some privileged access to the truth, or that everything can be reduced to one essential factor, whether it be the productive forces or something else. I have stressed the importance of production, and of class and class struggle, but even these factors are insufficient as an explanation for all developments in society. Therefore, in stressing struggle and/or agency over structure or the productive forces, I am not suggesting that everything in society is reducible to "the class struggle". This would be to replace one dogma with another. Engels (cited in Thompson 1965: 275) once complained that "the materialist conception of history . . . has a lot of friends nowadays to whom it serves as an excuse for *not* studying history . . . Our conception of history is above all a guide to study, not a lever for construction after the manner of the Hegelians." What Engels appears to be saying is that "historical materialism" is far more empirically open-ended than is often suggested, a point I return to in Chapter 4. On the other hand, the reason class is such a key category – but again, I stress, not necessarily the only category – in the study of modernity is that it provides an indispensable guide to understanding the *social* nature of the world. This is in contrast to so much social theory (and political practice), which is content to confine itself to the level of surface appearances.

This discussion leads directly to the content of the rest of the book, and especially the rest of Part I. Debate within the impasse in development studies has focused on the shortcomings of Marxist theory (Booth 1985). It will become clear however that there are strong grounds for arguing that these weaknesses belong to orthodox Marxism and its close cousin, structuralist Marxism. However, the relevance of this chapter to the rest of the book is not simply to show that Marx and orthodox Marxism are not necessarily one and the same thing. This chapter has attempted to show that orthodox Marxism, in presenting a general theory of history, actually fetishizes Marx's method. My argument in Chapters 3 and 4 is that such a fetishistic model has also pervaded development theory since 1945, and this is the basic reason for the impasse in development studies. It is in this sense that the next two chapters should be read as a *critique* of existing categories in development studies.

3

Modernization, dependency and development

This chapter examines the rise and fall of development theory since 1945. The first section examines the dominant discourse in development theory from the 1950s to the late 1960s, namely modernization theory. This section also includes an extension of the discussion of the previous chapter, and a brief reassessment of Marxist "modernization" theory. The second section examines alternatives to modernization theory and orthodox Marxism, namely underdevelopment, dependency and world systems theories. An assessment is then made, and in the third section I reintroduce one (orthodox) Marxist alternative to these theories, associated chiefly with the work of Bill Warren. In my conclusion I argue that there is a common thread in all these works, based on "developmentalism", which is best described as a post-war version of evolutionary theory.

Modernization theory

The theory in context: The Third World and development economics

Modernization theory arose in the historical context of the end of the European and Japanese empires and the beginning of the Cold War. It was in this context that the term "Third World" was first used by the independent left in France to describe a "third way" or "third path" between capitalism and communism (Thomas 1983, Worsley 1964, 1984). This notion was taken up by Third World leaders such as Nehru, Nasser and Sukharno, and they began to form organizations that challenged the primacy in the global order of the East–West conflict. The Non-Aligned Movement in particular argued for a political path in the Third World independent of the two superpowers. In challenging this perception, these leaders began to utilize the term Third World in a second sense, describing the inequalities in wealth and power between the rich North and poor South. In 1964, the United Nations Conference on Trade and Development (UNCTAD) was formed to challenge the inequities of the global order (Hoogvelt 1982: 74–80, Rajamoorthy 1993).

In many respects, non-alignment and the critique of the North–South divide were inspirations for at least some versions of dependency theory. The focus on the structural inequalities of the world system led to demands for controls on the behaviour of Western transnational companies and demands for fairer trading relations between North and South, which culminated in the mid-1970s' votes at the United Nations for a "new international economic order" (Anell & Nygren 1980: 187–209). Nevertheless, in the 1950s and 1960s these principles were in some ways compatible with modernization theory, despite some clear differences. This was the case for three reasons: first, non-alignment was often rhetorical from the start, and, in practice, by the 1980s most Third World countries openly associated themselves with one or the other superpower (Worsley 1984: 324–5, Halliday 1989: 21–2); secondly, both the Third World and modernization theory believed that the best way to achieve "modernity" was through economic growth, and especially industrialization; and thirdly, both perspectives criticized the neoclassical theory of international trade.

Modernization theory was comparatively moderate and far less critical of the global order than "structuralist" development economics. A leading structuralist economist, Raul Prebisch (1959: 261–4), argued that there was a tendency for the terms of trade of primary producers (concentrated in the Third World) to decline against manufacturing producers (concentrated in the First World). This focus on the international system was a great influence on dependency theory, and many dependency writers in the 1960s and 1970s came out of the United Nations Economic Commission on Latin America (ECLA), where Prebisch had previously worked. However, in the 1950s and early 1960s, his views were compatible with modernization theory because both schools accepted the strategy of economic growth through industrialization. For Prebisch, this was the best way for Third World countries to combat the tendency of their terms of trade to decline. For modernization theory, industrialization was the route for Third World countries to become modern. Both structuralism and modernization theory at least implicitly agreed that foreign capital would have to play the leading role in the early stages of industrialization (see Rostow 1971: 180, Lewis 1950: 38). This emphasis on industrialization represented a critique (albeit an implicit and limited one in the case of modernization theory) of the neoclassical theory of comparative advantage, whereby countries specialize in producing those goods in which they are most efficient, and then trade openly on the world market with the products of other countries' comparative advantages (see Ricardo 1971: 133–41, Ohlin 1933). Industrial development in the periphery meant the promotion of the production of goods that the Third World, initially at least, could not produce most efficiently or cheaply. In practice, this meant that industrial development was promoted on the basis of the protection of domestic industry from foreign competition, via a process of "import substitution" (Lewis 1950, Prebisch 1959, and, from within the modernization school, see Moore 1963: 97).

So, to summarize: modernization theory arose in the context of the rise of newly independent states and the Cold War between the capitalist and communist worlds. Although far more moderate than structuralist development economics, it shared some of its assumptions and most of the same strategic goals, particularly modernization through industrialization.

Modernization theory explained

The theory of modernization was an attempt by mainly First World scholars to explain the social reality of the "new states" of the Third World. The attainment of a modern society was seen as the strategic goal for these new nations, and this was defined as a social system based on achievement, universalism and individualism (Parsons et al. 1962: 76–88, see also Almond & Coleman 1960: 532–3). The modern, Western world of social mobility, equal opportunity, the rule of law and individual freedom was contrasted with traditional societies, which were based on ascribed status, hierarchy and personalized social relations. The purpose of modernization theory was to explain, and promote, the transition from traditional to modern society.

Modernization theory argued that this transition should be regarded as a process of traditional societies "catching up" with the modern world. As Moore (1963: 89) argued:

What is involved in modernization is a "total" transformation of a traditional or pre-modern society into the types of technology and associated social organization that characterize the "advanced", economically prosperous, and relatively politically stable nations of the western world.

The social psychologist David McClelland (1961: 430) similarly argued that the solution to backwardness was to educate the traditionalists on the virtues of the Western world, but he placed more emphasis on the role of Western ideas. However, modernization theory was most clearly elaborated by Walt Rostow (1960: 4–16), who argued that there were five stages of development through which all societies passed. These were:
  (i) the traditional stage;
 (ii) the preconditions for take-off;
(iii) take-off;
 (iv) the drive to maturity;
  (v) high mass consumption.
Third World societies were regarded as traditional, and so needed to develop to the second stage, and thus establish the preconditions for take-off. Rostow described these preconditions as the development of trade, the beginnings of rational, scientific ideas, and the emergence of an elite that reinvests rather than squanders its wealth (ibid.: 17–35). The theory argued that this process could be speeded up by the encouragement and diffusion of Western investment and ideas (Moore 1965: 10, Rostow 1971: 180).

Writers in this tradition also argued that industrialization would promote Western ideas of individualism, equality of opportunity and shared values, which in turn would reduce social unrest and class conflict (Kerr et al. 1962: chs 1 & 2). They argued that a similar process would occur in the Second World, too, as it developed through industrial growth (ibid.). This was because "a commercial-industrial system imposes certain organizational and institutional requirements not only on the economy but also on many other aspects of society" (Moore 1965: 11–12). It was therefore argued that modernization would promote convergence between the "three worlds".

Modernization theory was, therefore, an altogether more conservative theory than structuralist economics, even though it shared similar strategic goals. Indeed, given the Cold War context in which the theory was devised, it is at times unclear whether modernization theory was an analytical or prescriptive device, and it left a number of unanswered questions. For instance, the theory was unclear on the question of whether modernization was *actually occurring* or whether it *should occur*. It was also unclear as to the motives of those promoting modernization: was it to relieve poverty or to provide a bulwark against communism? Obviously the two factors are connected, but the subtitle of Rostow's classic ("a non-communist manifesto") suggests that the latter may have been considered more important than the former. I return to these questions in the critique below. For now, it should be clear that modernization theory was based on an evolutionary model of development, whereby all nation-states passed through broadly similar stages of development. In the context of the post-war world, it was considered imperative that the modern West should help to promote the transition to modernity in the traditional Third World.

Orthodox Marxism and modernization

"Orthodox" Marxism, as discussed in the previous chapter, shared an evolutionary, "stagist", approach to development. The previous chapter examined and criticized this *theory*, but I briefly return to it here to outline the *history* of orthodox Marxism and emphasize its similarities with modernization theory. It was shown in the previous chapter that, although in some respects Bolshevism challenged the historical materialism of the Second International, in other respects it replicated it. The Bolshevik seizure of power in 1917 was a challenge to the prevailing orthodoxy that the Russian social order was not "ripe" for socialist revolution, but, once in power, the Bolsheviks used the state to develop the productive forces. Socialism was defined less by its ability to emancipate labour and more by its ability to develop labour productivity more efficiently than capitalism. It was therefore seen as a different means to achieve the "modern" end of an industrial society based on mass production, and so was compatible with orthodox historical materialism, which saw history as the development of the productive forces.

As early as 1918 Lenin wrote (1977: 248) that "socialism calls for a conscious mass advance to greater productivity of labour compared with capitalism and on the basis achieved by capitalism. Socialism must achieve this advance in its own way . . . by Soviet methods. And the specialists . . . are, in the main, inevitably bourgeois." It was not surprising, therefore, that the ruling Communist Party was willing to utilize certain features of work organization in capitalist societies, especially the managerial strategy of Taylorism (see Littler 1984). After Lenin's death, the evolutionary version of historical materialism became official Soviet state dogma. Stalin wrote that "First the productive forces of society change and develop, and then, depending on these changes and in conformity with them, men's relations of production, their economic relations, change" (Stalin 1976: 859). On this basis, he identified "the laws of development of society" (ibid.: 848), in which nations pass through five stages of history: primitive communism, slavery, feudalism, capitalism and communism (ibid.: 862).

Later Soviet writers more or less repeated these statements (see the previous chapter) and largely glossed over the tensions in the work of Marx. For example, when the letters to Zasulich were discovered (see previous chapter), they were explained away by references to Marx's age and declining intellectual capacities (see Shanin 1984: 18–19). In terms of the post-war underdeveloped world, developments in Soviet Marxism did not stray far from the original paradigm of the primacy of the productive forces. The "non-capitalist path of development" was championed as a stage of development for Third World countries that would create "the material prerequisites . . . to the future socialist reconstruction of the economy" (Solodovnikov, cited in Lowy 1981: 197). Egypt, Somalia and Guinea, among others, were given Soviet support for their promotion of the "prerequisites for socialism", despite the fact that they were clearly capitalist societies. Although this theoretical development is obviously closely tied to the Cold War, the Soviet Union again "justified" the theory on the basis of the states' promotion of the productive forces (although this was hardly successful in some cases) (for a critique, see Thomas 1977).

Modernization theory assessed

Modernization theory has been subject to rigorous criticism by many writers (see Frank 1966, 1969a, 1969b, Bernstein 1971, Leys 1982, Webster 1990: 56–62, Spybey 1992: chs 1 & 2), although this has not prevented a partial revival of the theory, albeit in a different, often explicitly neoclassical form (see Little 1981: 25, Lal 1983: 45–7, Berger 1986, Fukuyama 1992). This revival is examined in later chapters. For now, the many criticisms that have been made of the theory will be reviewed. However, before doing so, I want to make a point that will become clearer in later chapters. While accepting much of the force of the criticisms made of this theory, I do not object to some of the key categories of the modernization school. My objections to modernization theory lie not with the concepts of "progress" or "modernity", but with the way that these concepts are employed. In other words, these concepts are useful if used in a flexible and non-evolutionary way. Indeed, without such concepts there could be no measure of whether a particular kind of development is desirable or not (see further, Chapter 7).

The basic problem with modernization theory is that it assumes that there is an unproblematic transition from traditional society to modernity. The cause of this transition is either Western individualism (entrepreneurship) or Western technology (the industrial society), or a mixture of these two factors. The problem with this approach is that the two factors are taken in isolation from the particular social structure in which they are embedded. Technology does not determine the social structure; rather it is the social structure that influences technology. This can be seen empirically through a comparison of different "industrial societies", which may utilize similar levels of technology but have radically different social and political structures (although this is not to deny some similarities as well). One could compare Japan, France and Britain, all relatively "advanced" industrial (or even "post-industrial") societies, but all with different forms of work organization, trade union structure, cultural relations and, to some extent, political systems (see Gallie 1983). In so far as there are similarities, these cannot be explained by the "determining role" of technology; instead, one would have to examine the specific social and political history of each country.

Such a technological determinism also leads to a rather naïve analysis of the reality of technologically advanced countries. The assumption that the USA was a modern meritocracy in the 1960s seems hopelessly wide of the mark when one considers the "racial" and ethnic antagonisms that existed at that time, and that still exist. It is simply a fantasy to assume that racial prejudice is a product of traditional antagonisms, which is bound to disappear through the modernizing drive of mass technology. Instead, what is required is a careful analysis of the history of so-called "race relations" in the United States, and therefore a far more balanced analysis of the tradition–modern dichotomy than modernization and convergence theories allow. In so doing, it might prove the case that "modernity" *creates* conflict, just as much as it may destroy it.

This point can be seen if one examines the Green Revolution in India in the 1960s. The aim of this revolution in agriculture was to increase agricultural productivity through a technological package which, it was assumed, would benefit all farmers. The result was an increase in productivity and output, but at the expense of increasing differentiation among the peasantry (see Desai 1975: 25–8, Byres & Crow 1988: 177–81, Byres 1989: 53–8). Harriss (1987: 237–8) has described why this differentiation occurred:

> those with more resources are able to make better use of the package than the small producers . . . [and] conditions are thereby created which are even more unfavourable to small producers (because the success of the relatively well off makes it more difficult for poorer cultivators to obtain labour, credit or fertilizers).

What this passage makes clear is that the enrichment of some, and impoverishment of others, occurred because technology was diffused into a highly unequal social structure, and so access to the technology was itself highly unequal. The result was an intensification, rather than alleviation, of inequality. Once again, this example shows that conflict may increase, rather than diminish, with the onset of "modernization".

Modernization theory conceptualizes the relationship between modernity and the world system in an equally naïve way. In so far as the theory examines relationships between different regions of the world, it assumes that these are based on a consensus and shared values, rather than on unequal power relations or conflict. Obstacles to modernity exist within a nation-state, not beyond it. But this thesis is based on a complete separation, or *dualism*, between "traditional" nations and the modern global order. As Frank (1969b: 62) has argued, such theories "are inadequate because the supposed structural duality is contrary to both historical and contemporary reality: the entire social fabric of the underdeveloped countries has long since been penetrated and transformed by, and integrated into, the world embracing system of which it is an integral part".

Modernization theory therefore separates the modern international division of labour from "traditional" nation-states, and fails to clarify the mechanisms by which the latter were integrated into the former. So, for example, New World slavery should not in any sense be regarded as "traditional" because it was the product of the will of specifically modern powers (E. Williams 1987). This example again suggests that the reality of modernity is far more complex, and conflict ridden, than modernization theory allows. In peripheral societies today, such conflict may rest on a "dualism" not only between First and Third World, but also within Third World societies themselves. For example, so-called modern sectors may exploit so-called traditional sectors through access to cheap labour or raw materials produced in the latter.

Modernization theory therefore rests on the assumptions of a conflict-free modern world, a rigid dichotomy between tradition and modernity, and an unproblematic transition from one to the other. It is the third of these problems that is the central weakness of orthodox Marxism, which rests on a view of history whereby the productive forces develop, and relations of production evolve, according to their functionality to the continued development of the productive forces. The problem with this thesis, as I have already argued, is that it assumes that the class that can best develop the productive forces has the capacity to change the mode of production to ensure the continued development of the productive forces (see Levine & Wright 1980: 66). Both theories therefore share a technological determinism that fails to locate technology in a particular social structure (or, if one prefers, the relations of production).

It is ironic then, given the Cold War context in which they operated, that modernization theory and orthodox Marxism actually shared similar evolutionary assumptions. It may be true that both versions of modernization theory were more important as justifications for respective superpower interventions in the periphery, rather than the actual coherence of the theories themselves. Nevertheless, conclusions such as this can be drawn only after, and not prior to, analysis of the logical consistency of a particular theory. Both were criticized on these grounds by an alternative school of thought, which developed in the late 1960s and to which I now turn.

Underdevelopment and dependency theory

Underdevelopment and world systems theory

This school of thought is associated with the work of Andre Gunder Frank, but the influence of Paul Baran's work (1957) is also very important. He argued that the economic relationships that existed between western Europe (and later Japan and the United States) and the rest of the world were based on conflict and exploitation. The former took part in "outright plunder or in plunder thinly veiled as trade, seizing and removing tremendous wealth from the place of their penetration" (Baran 1957: 141–2). The result was a transfer of wealth from the latter to the former (ibid.: 22–4).

Baran's work is more complex than this but it was his theory of surplus transfer that most influenced Frank. Frank criticized the dualist thesis (see above), which isolated "modern" and "traditional" states, and argued that the two were closely linked. He applied his critique to both modernization theory and orthodox Marxism (1969b: 238), and instead argued, contrary to Baran, that the world had been capitalist since the sixteenth century (Frank 1969a: 14–15, for Baran's position, see Baran 1957: 164). The colonial period was characterized by the placing of regions into "a situation of growing subjection and economic dependence, both colonial and neo-colonial, in the single world system of expanding commercial capitalism" (Frank 1972: 13).

Frank argues that the capitalist world is based on a chain of metropolis–satellite relations, divided by nations and regions within nations. The metropoles exploit the satellites through the expropriation of economic surplus, which the former use for their own economic development. The satellites, meanwhile, "remain underdeveloped for the lack of access to their own surplus" (Frank 1969a: 9). The dualism of modernization theory and orthodox Marxism is replaced by a theory that argues that the world has been capitalist since the sixteenth century, and this is because all sectors are drawn into the world system, based on production for the market. The result is that some nations develop at the expense of others; as Frank (1969a: 240) argues, "[t]his capitalist system has at all times and in all places – as in its nature it must – produced both development and underdevelopment. The one is as much the product of the system, is just as 'capitalist' as the other."

Frank's central argument is therefore quite simple. Modernization theory and orthodox Marxism counterposed the modern world to the traditional world, and gave their support to those that promoted the modern, but in fact it is actually these "modernists" that have created the "traditional" world. "Traditional" sectors – or satellites – are as much a feature of the world system as "modern" ones – or metropoles. Metropoles exploit the satellites by expropriating their surplus, and so the two sectors must not be seen in isolation, but are in fact part of the same structured whole.

Wallerstein developed this idea of the world capitalist economy in his "world system analysis". His basic argument was that the creation of the world capitalist economy in the sixteenth century led to a new period of history, based on expanded accumulation rather than stagnant consumption. This was attributable to the emergence of three key factors:

an expansion of the geographical size of the world in question (incorporation), the development of variegated methods of labour control for different products and different zones of the world economy (specialisation) and the creation of relatively strong state machineries in what would be the core states of this capitalist world economy (to assure transfer of surplus to the core). (Wallerstein 1974: 7)

According to Wallerstein then, the growth of trade, the more efficient organization of production, and a structure of unequal nation-states led to the creation of the world capitalist economy. This in turn led to an unprecedented expansion of the productive forces, but at the expense of the periphery (Frank's satellites) and to the benefit of the core (Frank's metropoles).

Further developments – unequal exchange and dependency theory

The work of Arghiri Emmanuel and Samir Amin complemented that of Frank and Wallerstein because they attempted to clarify some of the mechanisms whereby surplus was transferred to the core from the periphery. Their respective work has been summarized and criticized elsewhere (Bettelheim 1972, Pilling 1973, Brewer 1980), so I will be brief. Emmanuel's theory (1972) is very similar to Raul Prebisch's in that he argues that surplus is transferred to the core from the periphery through a process of unequal exchange. He assumes that the rate of surplus value expropriated by capital is greater in the periphery than in the core, but that there is also an equalization of the rate of profit in the world economy. This leads to a transfer of surplus value from the Third World to the First World.

In essence then, Emmanuel is arguing that wage discrepancies between workers in core and periphery lead to a process of unequal exchange and surplus transfer (1972: 43). Amin (1976: 288) essentially concurs with this analysis, although his analysis also examines the ways in which the periphery's incorporation into the world economy led to its "extroverted" or "export-oriented" accumulation. According to Amin (1976: 288), the dominance of the periphery by the centre promotes three key distortions in the former:

(1) a crucial distortion toward export activities, which absorb the major part of the capital arriving from the centre;
(2) a distortion toward tertiary activities, which arises both from the special contradictions of peripheral capitalism and from the original structures of the peripheral formations; and
(3) a distortion in the choice of branches of industry, toward light branches, together with the utilisation of modern techniques in these branches.

This pattern of export activity leads to a vicious circle of underdevelopment, which led Amin (1976: 13) to conclude that "so long as an underdeveloped country continues to be integrated in the world market, it remains helpless . . . the possibilities of local accumulation are nil".

The work of Frank, Wallerstein, Emmanuel and Amin came under increasing criticism in the 1970s. These criticisms are examined in detail below, but one major reason was the logical implication of the thesis that the contemporary Third World would be stagnant, and development would be impossible, so long as nations in the periphery remained tied to the core-dominated world system. The problem for the theory was that it hardly fitted the reality of high growth in much of the Third World in the 1960s and 1970s. All kinds of contorted attempts were made to explain this problem (see below and Chapter 5), but most writers accepted that Frank's stagnationist thesis was too strong. One response by those sympathetic to his focus on the international economy was to introduce a "softer" version of underdevelopment theory, known as dependency theory. This approach was also influenced by the work of Prebisch, but it argued that his solution to the problem of unfavourable terms of trade – that is, industrialization – had led to a new one – dependence. The Latin American economist Theotonio dos Santos (cited in O'Brien 1975: 12) summarized the basic hypothesis:

> Dependence is a conditioning situation in which the economies of one group of countries are conditioned by the development and expansion of others.

First World countries may advance through self-impulsion, but the dependent economy can "only expand as a reflection of the expansion of the dominant countries, which may have positive or negative effects on their immediate development" (ibid.).

So, in arguing that development could occur in the periphery, dependency theory challenged a central argument of underdevelopment theory. On the other hand, in arguing that this peripheral or dependent development was largely at the behest of the core, the theory retained another of Frank's central arguments (see also Cardoso 1982, Cardoso & Faletto 1979).

A critique

These theories have all been challenged by both neoclassical and Marxist theories. The industrialization of parts of the periphery led to a return to modernization theory, although this time it was more clearly associated with neoliberal economics (Little 1981, Berger 1986, Fukuyama 1992), and to a return to orthodox Marxism, though this too was often in an amended form (Warren 1973, 1980). The contentions of these theories are analyzed in more detail in the third section, and detailed debates around the industrialization of the Third World are discussed in Chapter 5. What I want to do here is focus on the methodological problems of these theories, and draw out some of their implications. My critique of underdevelopment theory and its close relatives is divided into five closely related parts: (a) its analysis of trade; (b) its analysis of class; (c) its methodological approach to core and periphery; (d) its approach to evolutionary theory; (e) its political implications.

Underdevelopment theory and trade
The central contention of underdevelopment theory is that the core develops through the underdevelopment of the periphery, which takes place through a process of the former extracting the surplus of the latter. But, as the neoclassical economist Peter Bauer (1984a: 134) points out, the poorest nations of the world are precisely those nations that trade only in small amounts. The force of this argument is powerful, even if one need not accept Bauer's "medicine" for alleviating the situation. In fact, both neoclassical and underdevelopment theory, in focusing on trade relations, neglect an analysis of how a surplus is produced in the first place. This can be illustrated through a discussion of underdevelopment theory's account of surplus extraction.

The precise mechanisms of this process are not always made clear, but trade obviously plays a key role. However, this leads to two questions: what are the origins of capitalism? and why do some countries benefit, and others lose, within this world economy? The implication – and this seems to be confirmed in the quotation from Wallerstein above (and in his most explicitly neoclassical position in Wallerstein 1986) – is that capitalism has its roots in the emergence of individuals who have the motivation to pursue profit maximization. The emergence of these entrepreneurs in western Europe enabled this region to grow at the expense of the rest of the world. The problem with this contention is that individual profit maximizing takes place within a particular social structure, which (unlike in Marx's work) is never emphasized by this school of thought. In so far as the social structure is analyzed, it is explained as a product of the needs of the system – that is, profit-making entrepreneurship creates a social structure and division of labour, rather than vice versa. So, to refer back to Chapter

2, the underdevelopment school lacks an adequate conception of the origins of capitalism and so-called "primitive accumulation" (Marx 1976b: 873).

An example should illustrate my point. Surplus extraction is hardly unique to capitalism; as Dore & Weeks (1979: 65) state, "it is no more characteristic of capitalism in its imperialist stage than it was characteristic of competitive capitalism or ancient Rome". This point is accepted by Wallerstein (1974: 38). But this begs a further question, which has been asked by Brenner (1977: 67):

> what allowed for, and ensured, that wealth brought into the core from the periphery would be used for productive rather than non-productive purposes?

The implication is clear that – for this theory at least – the only explanation can lie in entrepreneurial, profit-making activity. It is ironic, therefore, that this theory, which is diametrically opposed to the neoliberal view of comparative advantage, actually shares the same starting point of profit-making individuals. Brenner (ibid.: 58) makes this point clear when he argues that the model assumes "the extra historical universe of *homo oeconomicus*, of individual profit maximizers competing on the market, outside of any system of social relations of exploitation". Marx (1976b: 739) criticized classical political economy on exactly these grounds, arguing that the capitalist "shares with the miser an absolute drive for self-enrichment. But what appears in the miser as the mania of an individual is in the capitalist the effect of a social mechanism in which he is merely a cog" (see also on Wallerstein, Bergesen 1990: 74, and the Dobb–Sweezy debate in Hilton 1976a).

The implications of this focus on exchange relations can also be seen if one examines Emmanuel's account of unequal exchange. His starting point of different rates of surplus value combined with an equalization of the rate of profit leads to the inevitable conclusion that either there is no need for uneven development to occur in the international economy, or capitalist development should be occurring most rapidly in the Third World, where the rate of exploitation is higher (see Dore & Weeks 1979: 71). Either way, this suggests that the neoliberal theory of comparative advantage is right after all.

It is for these reasons that underdevelopment theory has been described as a "neo-Smithian" Marxism (Brenner 1977, Jenkins 1984a). Its contentions are diametrically opposed to the neoclassical theory of comparative advantage, but in fact it shares a very similar starting point. This is because its analysis of the social structure is not well grounded and, as a result, it assumes capitalist behaviour independently of capitalist relations of production.

Underdevelopment theory and class analysis
Given that this theory takes trade as its starting point, it is not surprising that its analysis of class is derived from its functionality to the world capitalist market. Frank's analysis (1969b: 340–9) is based on spatial categories in which some people benefit at the expense of others in the network of metropolis–satellite relations. Wallerstein (1974: 127, see also Wallerstein 1983: 18) has similarly argued that:

> the relations of production that define a whole system are the "relations of production" of the whole system and the system at this point in time is the European world economy. Free labour is indeed a defining feature of capitalism, but not free labour throughout the productive enterprises . . . free labour is the form of labour control used for skilled work in core countries whereas coerced labour is used for less skilled work in peripheral areas. The combination thereof is the essence of capitalism.

Leaving aside the (highly questionable) empirical validity of this assertion, Wallerstein's basic (and, once again, neo-Smithian) contention is that the class structure is a product of the functional requirements of the world system. But class structure is never predetermined and is the result of human agency, and it should not be assumed that it will automatically "fit the logic" of the needs of the world system (see Brenner 1977, Connell 1984).

Underdevelopment theory and methodology
The problems discussed above can be traced to the methodology used by this theory. It is a theory that hardly examines, nor needs to examine, the periphery. The periphery is homogenized as an exploited mass (Leys 1982), and changes in it are read off from the needs of the core (Booth 1985, Corbridge 1986). The assumption that the underdeveloped world can in some ways develop and still remain part of the

world capitalist system is excluded on *a priori* grounds. When there are changes, such as rapid industrialization in east Asia, this is dismissed as a consequence of the needs of the metropolis, as not being genuine capitalist development, or as a change in status for some countries within the omnipresent world system (see Frank 1983). However to argue that, for example, east Asian newly industrializing countries have advanced from periphery to semi-periphery (Caporaso 1982) is hardly to explain it, still less does it explain how this industrialization might itself change the world system itself. Instead we are left with a static model of a world system in which changes in it, and especially changes in the periphery, are assumed to be "functional" to the system's needs.

Even the impressive work of Cardoso can be criticized on similar grounds. Although recognizing that dependency does not preclude industrial development, he still argues that it is in some sense dependent. The problem with this view is that he still begins with the *idea* of dependency, rather than examining the forms of foreign penetration (Cardoso 1987) and the impact of this on the class structure, and then asking whether this leads to dependence (see, on the Korean case, Hamilton 1986: 115). This suggests that dependence may be more useful as a concrete, *a posteriori* concept, rather than an abstract, *a priori* theory (Palma 1978: 911–2).

Nevertheless, despite this weakness, Cardoso's work remains impressive, because it attempts to examine the specificity of class formation and capital accumulation in parts of Latin America. In this respect Cardoso's work presents a useful example of the way forward for "post-impasse" development theory because it recognizes the structural constraints faced by specific Third World nations, but does not fetishize these constraints so that they are seen as insurmountable and change is regarded as automatically serving the functions of global capital (Cardoso 1987: 13, see also Chapter 5).

Underdevelopment and evolutionary theory
One of the great strengths of this school of thought is its analysis of the relations of conflict that exist in the international political economy. In focusing on these relations, underdevelopment theory undermined evolutionary models, which saw societies (defined as nation-states) passing through similar stages of development. This critique was accomplished by arguing that the incorporation of

traditional societies into the world economy was a specifically modern phenomenon.

However, this challenge to evolutionary models was itself based on an implicit acceptance of its basic premises. "Genuine" capitalist development is ruled out by this theory, but this can be done only by having an implicit "model" of what constitutes "normal" capitalist development. In this sense the theory is a "mirror image" of evolutionary theories, where "inevitable development" in the case of modernization is replaced by "inevitable stagnation or distortion" in underdevelopment and dependency (Phillips 1977, Bernstein 1979, 1982: 227, Spybey 1992: 32). It also leads to a conception of socialism as "national development" (Bernstein 1982: 227), which is not unlike the conception of socialism in the orthodox Marxist model.

Underdevelopment theory and politics
This conception of socialism as genuine national development can be seen in the work of Amin and Frank, who both emphasize the need for autarky for genuine Third World development to occur (Amin et al. 1982: 225). This is not surprising when the focus is on how trade underdevelops the periphery; the logical strategy becomes one of delinking from the world economy. Although Frank was initially optimistic about the prospects for such a strategy, by the early 1980s he had become a pessimist, arguing that "this theoretical alternative never existed in fact" (Frank 1982a: 135). Such an excessive pessimism is the logical corollary of Frank's excessive optimism in the late 1960s, which was linked to his essentially voluntarist account of social revolution. In both accounts the theory neglects "the dynamics of class and popular struggle within UDCs (underdeveloped countries), and the specific contradictions they express" (Bernstein & Nicholas 1983: 611, see also Petras & Brill 1985, Gulalp 1987).

There are also more immediate political concerns that can be derived from underdevelopment theory. For Frank to remain consistent to his theory, he should support the US trade boycott of Cuba, or the decline in real terms of British aid to the Third World, as good examples of "anti-imperialist action". The emphasis on autarky, or delinking from the world economy, has led to a neglect of struggles within the Third World, and an approach to some repressive Third World governments that is too uncritical. Frank and Amin's policy of "delinking" from the world economy and support for "self reliant socialism" led to a rather uncritical approach to the Khmer Rouge in Cambodia (see Amin et al. 1982: 218). Other writers who share a similar methodology have emphasized how the US bombing of Cambodia in the late 1960s helped to pave the way for Pol Pot's accession to power. I do not wish to deny that the United States' government must accept major responsibility when it carried out these illegal and murderous acts, but it is stretching a case to argue that "the US created Pol Pot". This argument, like the methodology of underdevelopment theory, simply denies the ability of Third World peoples for meaningful action (for Wallerstein, such struggles are simply irrelevant – see Wallerstein 1983: 34–5), and it therefore ironically replicates the imperialism that it so determinedly criticizes.

The return to orthodox Marxism: the work of Bill Warren

The work of Bill Warren arose in the context of the industrialization of a number of Third World countries, which, he argued undermined the central "stagnationist" claim of underdevelopment theories (1973). This topic is the subject of Chapter 5, but here I want to summarize and criticize Warren's methodological approach. As this largely represents a return to one particular version of orthodox Marxism, I will be brief.

Warren (1980: ch.7) argues that socialists have been misled by the nationalist mythology of dependency theory and supported Third World nationalism at the expense of international socialism. The view that imperialism retards the development of the productive forces is (at least in his 1980 work) traced back to Lenin, who described capitalism in its monopoly stage as parasitic and in a state of decay (ibid.: chs 4 & 5). In fact, Warren argues, imperialism has promoted the development of capitalism in the Third World, and the obstacles that exist to its further development must be rooted firmly in the internal conditions of Third World societies (ibid.: 253).

So, like some of Marx's writings on colonialism (see previous chapter), Warren supports capitalism in the periphery and argues that imperialism at least provides the basis for expanding this progressive development. He argues (Warren 1973: 41, see also Warren 1980: 136) that:

> If the extension of capitalism into non-capitalist areas of the world created an international system of inequality and exploitation called imperialism, it simultaneously created the conditions for the destruction of this system by the spread of capitalist social relations and productive forces throughout the non-capitalist world. Such has been our thesis, as it was the thesis of Marx, Lenin, Luxemburg and Bukharin.

Capitalism, and by implication imperialism, is seen as progressive because it promotes the development of the productive forces, which sows the seeds of socialism (Warren 1980: 11–18). It does this not only by a process of the material development of the productive forces, but by the destruction of traditional cultures (ibid.: 18–25), which also serves as a bridge to socialism. The fascism of two relatively advanced capitalist states in the 1920s and 1930s is dismissed as "an interlude that arose not from fundamental characteristics of capitalism as such . . . but from specific conditions in particular states" (ibid.: 28).

Clearly, then, Warren is adamant that there is a "developmental logic" to capitalism, which appears to occur regardless of the actions of human beings – except, of course, when their actions do not conform to such a logic. He argues that "[c]apitalism and democracy are . . . linked virtually as Siamese twins", but adds as a footnote that this observation relates specifically to western Europe (ibid.: 28). The clear assumption, in classic Rostowian fashion, is that, because it happened in the West, then it must happen elsewhere. The rise of democracy in Europe is thereby "explained" as the product of the logic of capitalism rather than as the product of the *struggles* of exploited classes in capitalist society. Once again, a grand theoretical model fetishizing the real world is used instead of an analysis rooted in the history of real social struggles.

For all of Warren's criticisms of the orthodox Marxism of the Comintern (the Third International), his work should be located in precisely this paradigm. The Third International saw imperialism as a hindrance to the development of the productive forces, while Warren sees it as a promoter of this development. Both are searching for an agency to carry out a similar developmental logic. This preoccupation with development of the productive forces:

leads Warren to an abstract vision of capitalism and, further, to an extension of this vision to that of imperialism. His arguments are couched within a framework that excludes the actual mechanisms of conflict between classes, as well as capitals. (Gulalp 1986: 149)

Warren therefore abstracts from the class forces that exist in a particular space and at a particular time. Western capitalism – but not the struggles that created Western capitalism – is upheld as a model for the rest of the world so that the process of history, the development of the productive forces, can continue. Rostow's tradition–modern dichotomy, and along with it the inability to explain (among others) the creation of slavery in the New World and forced labour in the colonies, has returned. It seems that, for Warren, there is only one form of capitalist development.

This developmental approach can also be seen in the work of a number of Warren's sympathizers. For example, John Sender and Sheila Smith (1985, 1986, 1990) have convincingly challenged some of the key contentions of underdevelopment theories. In particular they have shown (1986: 113–27, see also Warren 1980: 140–3) that the economic performance of African countries cannot simply be "read off" from an alleged tendency for the terms of trade to decline, because this varies between African countries with similar export patterns. However, even this point should be qualified. Although these writers are undoubtedly correct in arguing that the problems of the Third World cannot be explained solely by external causes, they fail to prove that internal factors are the sole cause of "backwardness". The fact that the problems of the Third World cannot be read off from unfavourable terms of trade does not mean that these factors cease to exist (Larrain 1989: 197, see also Spraos 1983, which, contrary to Sender and Smith's common implication, argues – albeit in a far less mechanistic way than most writers – that the terms of trade are often unfavourable for primary producers).

Moreover, the developmental logic of these writers leads them to some odd prescriptions for development of the productive forces. It is argued that the further development of the productive forces can be achieved by the promotion of capitalist relations of production (Sender & Smith 1986, also Mueller 1980, Hyden 1983). The problem with this approach is that it again rests on a model of capitalism that is abstracted from real historical processes – and, in the case of

Africa, from the emergence of settler capitalism in colonial Kenya, Rhodesia and South Africa (G. Williams 1987: 650). Sender and Smith recognize that there will be costs in developing capitalism (for example 1986: 77), but they want to have it both ways, by supporting the strengthening of progressive class forces and by placing "the issues of trade union rights, wages and working conditions on the agenda" (ibid.: 132). How this concern fits with their "success stories" of Malawi, Kenya and the Ivory Coast is unclear, but what is more relevant here is how their technicist approach to development and the productive forces leads to a purely technicist account of the state and policy-making. They rightly argue that successful capitalist development rests on the establishment of a capitalist state. However, they then argue that "a method must be devised for the appropriation of sufficient surplus to ensure the smooth functioning of the military and repressive apparatus", as well as policy measures such as higher prices paid to farmers to maintain food supply (1986: 112). While there is nothing necessarily wrong with this particular policy prescription, it is wishful thinking to imply that Africa's problems are *simply* the result of incorrect policies. Once again, the "rational state" of advanced capitalist societies is transformed into an abstract, ahistorical model, which deserves support because it will further the development of the productive forces. The "rational entrepreneur" of neoclassical and underdevelopment theory is replaced by the "rational state", which is divorced from the existing social structure. If one analyzes the existing social structure, rather than a supposedly future one based on Western capitalism as a model, then it is possible to conceptualize the "irrational state" based on corruption and patronage as "a highly rational and *modern* attempt by capitalistic interests to appropriate monopoly rents through privileged access to officialdom" (Brett 1988: 28–9). The struggle to overcome these factors is a social struggle, not simply a technical question of ensuring "the smooth functioning of the military and repressive apparatus".

So, the work of Warren and his sympathizers is best characterized as a return to the stagism of orthodox Marxism. Social struggle is subordinated to the implementation of abstract models of a "normal" capitalism and "ideal" state, which develop the productive forces and therefore provide the material conditions for a future socialism. Socialism is rejected as inappropriate for backward societies (Sender & Smith 1990: 132) and capitalism is supported because it is best able to develop the productive forces. This approach rests on an ahistorical conception of capitalism, a neglect of international forces (see for example Sender & Smith 1986: 11–13), and a "fetishism" of the state (Bernstein 1987: 169–70).

Conclusion

Modernization theory, underdevelopment theory and the orthodox Marxist theory of the primacy of the productive forces have all been rejected in this chapter. Although these theories were counterposed to each other, they actually share three critical weaknesses: first, they each have a "developmental" logic; secondly, they each read off changes in the periphery from the needs of the core; and thirdly, they share a similar conception of progress.

Developmentalism

This can be defined as the idea that people are "passive actors" in the grand theatre of history, and that social change has a predetermined logic, which moves society towards a certain end (for a critique of this evolutionism, see Giddens 1984: 236–9). In their obsession with fetishized theoretical models, these theories have lost sight of "those social forces and formations which by their rise and fall, by their dissolution and recomposition largely shape and make history" (Tenbruck 1990: 205).

Thus, debates between underdevelopment and orthodox Marxist theories have focused on *capitalism in the abstract* and its capacity (or lack of it) to bring about development. The result is the "impossibilism" of Frank et al. against the "inevitabilism" of Warren et al. (and modernization theory). As Gulalp (1986: 155) has argued, "the terms of the debate are couched within a framework that idealizes the experience of western capitalist development and asks whether or not the idealized experience is reproduced in the rest of the world" (see also Williams 1978a). Development is thereby naturalized, rather than analyzed as a social and historical process.

Core–periphery relations

This idealization of the Western experience leads on to the second problem, which is that both sides examine the periphery only in so far as it is determined by the centre. Underdevelopment theories see the centre as an obstacle to its developmental logic, while modernization theories see the centre as the promoter of a developmental logic. Orthodox Marxism falls on either one side or the other; imperialism either retards the development of the productive forces (the Third International) or it promotes them (Warren, Sender and Smith). Neither side is prepared to account for the historical specificities of different Third World countries, which cannot be read off from an imperialism that is assumed to be homogeneous in its effects on the periphery. As Gulalp (1986: 156–7) states:

> imperialism . . . is neither an agent of diffusing the capitalist mode nor that of creating underdevelopment as an abstract process with a singular outcome. It is rather the process of creating an international division of labour with different outcomes in different parts of the world as well as in different phases of the development of capitalism on a world scale.

This has implications for how one conceptualizes the global political economy and the place of the Third World in it, which I return to in later chapters. For now, it should be clear that these theories rest on the common methodological weakness of seeing the "rest" in the image of the "West", which in turn can be derived from a common evolutionary position on development.

Progress

The principal consideration of each side in the debate is therefore economic growth. Progress is defined in terms of "the expansion of the market" or the "development of the productive forces" (Williams 1978a: 925). But this conception of progress is based on only one side of modernity and fails to capture its "dark side". Social actors thereby become subordinated to the "inevitability of progress", with all the social costs that this entails.

This is most clear in the case of the environment, where it has become increasingly obvious that it is simply impossible for all nation-states to follow similar paths of development for the simple reason that it could not be environmentally sustainable. As Gandhi (cited in T. Allen 1992: 389) commented in 1928:

> God forbid that India should ever take to industrialization after the manner of the West. The economic imperialism of a single tiny island kingdom is today keeping the world in chains. If an entire nation of 300 million took to similar economic exploitation, it would strip the world bare like locusts.

So, the "contribution" of these theories to the impasse is that they are based on an evolutionary logic that theorizes a model of ideal capitalist development, that has the same effect, both in different places and at different times. The next chapter examines one response to this problem.

4

Structuralist Marxism and development

A renewed interest from the late 1960s in "unorthodox" approaches to Marxism paved the way for an alternative to mainstream development and underdevelopment theories in the 1970s. Modernization theory was criticized for its evolutionary approach, while underdevelopment theory was rejected because it focused on trade relations. These theories were contrasted to a "scientific" method that took the mode of production as the determining concept. Although there were some divisions within this school of thought, the most influential thesis argued that Third World social formations were characterized by an articulation of modes of production.

This chapter provides a critical analysis of the modes of production school, and assesses some of the suggested alternatives to it. In making this assessment, I enter into a direct dialogue with those writers who have identified the impasse in development theory. In doing so, I argue that their identification of an impasse is essentially correct, but I do not share all of their arguments concerning the way out of it. This will also become clearer in the second part of the book.

The chapter is divided into three sections. First, I examine the theory of articulation of modes of production. Then I provide a critique of this theory, and show its similarities with the development theories discussed in the previous chapter. I extend the critique in the third section, which discusses the relationship between capitalism, unfree labour and the state, which in turn leads me back to an examination of the impasse, and especially of the relationship between theory and method. It is at this point that I return to the impasse, and express some reservations about some of the "post-Marxist alternatives" to the impasse.

The articulation of modes of production

Modes of production theory attempted to overcome the basic problems of the development theories discussed in the previous chapter. Modernization theory and orthodox Marxism were rejected for their "evolutionism" and their monocausal explanations, while underdevelopment theory was rejected for its "mirror image" of evolutionism and its focus on trade relations rather than the relations of production (see Laclau 1971/1977, Taylor 1979). In terms of its challenge to evolutionary theory, the new approach was an attempt to move "beyond the sociology of development" (Oxaal et al. 1975).

The theory emphasizes that capitalist development differs in both time and space, and so rejects models that argue that it uniformly develops or underdevelops the periphery. It is also argued by this school of thought that an emphasis on the determining role of the mode of production leads to a greater understanding of the mechanisms of "underdevelopment" (or, more precisely, uneven development). This is said to be the case because trade relations presuppose what they are supposed to explain – the production of a surplus product and the division of the world into core and periphery (see Laclau 1977, Dore & Weeks 1979: 65). The mode of production, on the other hand, takes as its point of departure the production of the surplus product and so is able to move to an explanation of the division of the world between core and periphery based on modes of production rather than trade relations *per se* (Mandle 1972: 53). The core therefore coincides with the capitalist regions of the world, which are largely based on free wage labour (see Chapter 2), while the periphery was incorporated into the world economy on the basis of unfree relations of production (that is, non-capitalist modes of production), which prevented an unprecedented accumulation of capital. Unequal trade relations were therefore not unequal trade relations *per se*, but in fact reflected unequal relations of production. It is for this reason that the "advanced" capitalist countries were able to dominate other areas of the world where non-capitalist modes of production existed (Bettelheim 1972).

Modes of production theory does however recognize that this relation is not static, and that capitalist relations of production have emerged in the periphery. Nevertheless, it is a *specific kind* of capitalism, one that is qualitatively different from its form in the core countries. Capitalism in the periphery is different because it is combined with non-capitalist modes of production – in other words, capitalism coexists, or "articulates", with non-capitalist modes (see Post 1978: 27). Non-capitalist production may be restructured by imperialist (that is, "core capitalist') penetration but it is also subordinated to it by its very "conservation" (see Bettelheim 1972).

At the heart of the theory is a distinction between social formation and mode of production. The social formation refers to a combination of economic, political and ideological practices or "levels". The mode of production refers to the economic level, which determines which of the different levels is dominant in the "structured totality" that constitutes the social formation. The economic level sets limits on the other levels, that carry out functions which necessarily reproduce the (economic) mode of production. These non-economic levels therefore enjoy only a relative autonomy from the mode of production (see Althusser & Balibar 1979: 178–80, and glossary, 319).

Within the periphery, any social formation (or "economic system" – see Laclau 1977) may be constituted by more than one mode of production (Taylor 1979: 105–42, Wolpe 1980: 19–27). Such an articulation does not lead to a new version of dualism because the different modes of production are closely linked within this social formation. Thus, for Laclau (1977: 35) an economic system is based on "the mutual relations between the different sectors of the economy, or between different production units, whether on a regional, national or world scale" and can include "as constitutive elements, different modes of production". Thus in an economic system the capitalist mode of production may be dominant and other modes of production are subordinated to it. Therefore, where the capitalist mode is dominant, the preservation of non-capitalist modes is explained by reference to the needs, or *reproductive requirements*, of the dominant capitalist mode. As Bradby (1975: 129) states:

Capitalism has different needs of pre-capitalist economies at different stages of development, which arise from specific historical circumstances, e.g. raw materials, land, labour power, and at times of crisis, markets.

Although some proponents of this theory also analyze the "laws of motion" of the non-capitalist mode (especially Rey, see Foster Carter

63

1978), the main focus of this school is on articulation where the capitalist mode dominates, and so the needs of the capitalist mode are paramount. Some examples illustrate their argument. In the case of various agricultural modes of production, the "peasant-worker's" continued access to land enables the capitalist to pay a wage lower than he or she might otherwise have to, and the peasant is exploited by the capitalist who buys the peasant's produce cheaply and sells it more expensively in a wider market place (Cliffe 1982). In the case of the so-called "informal sector", a "petty commodity mode of production" serves the requirements of the capitalist mode by providing a reserve army of labour, and by providing cheap goods to workers in the capitalist sector, which in turn reduces the reproduction costs of workers, thus allowing capitalists to again pay cheaper wages than would otherwise be the case (Moser 1978).

Perhaps the most interesting example, though, is provided by the work of Claude Meillassoux. He argues that imperialism conserves the "domestic mode of production" as a means of preserving high imperialist profits through the exploitation of female labour, which carries out domestic tasks. As he states (1981: 95):

> It is by establishing organic relations between capitalist and domestic economies that imperialism set up the mechanism of reproducing cheap labour power to its profit – a reproductive process which, at present, is the fundamental cause of underdevelopment at one end and of the wealth of the capitalist sector at the other.

Wolpe (1972; see also Wolpe 1980 and 1988) has utilized this theory to explain a specific form of capitalism in what became South Africa. Instead of an unambiguous break-up of non-capitalist modes in the late nineteenth century, the African labour force was dispossessed from its land (see Bundy 1979), but was effectively relocated on to 13 per cent of the land. The result was that the labour force working in the mines still had direct access to the means of production, which enabled capitalists to pay workers lower wages than would have been the case in a "free labour market". As Wolpe (1972: 429) argues:

> The state has been utilized at all times to secure and develop the capitalist mode of production. Viewed from this standpoint, racist

ideology and policy and the state now not only appear as the means for the reproduction of segregation and racial discrimination generally, but also as what they really are, the means for the reproduction of a particular mode of production.

Modes of production theory is also often closely linked to the notion that capital has become more international, as it crosses the boundaries of different nation-states. An articulation may occur in that "national capital" articulates with international capital, and vice versa, or international capital may have the effect of conserving non-capitalist modes within Third World social formations. In either case, it is assumed that the interests of capital will be realized (see Palloix 1975, Cypher 1979, Browett 1985).

So, proponents of the theory of articulation of modes of production argue that capitalism in the periphery "coexists" or "joins together" with non-capitalist modes of production to form a social formation. "Underdevelopment" is therefore not a product of trade relations but a result of the preservation of the non-capitalist mode of production. This preservation prevents the general emergence of free wage labour and therefore generalized commodity production, and so the development of the productive forces is not as rapid as in those (core) countries (or regions) where the capital–labour relation is generalized. The two modes are not however "separated" (as modernization theory contends), but are closely linked to the limits of the pre-capitalist mode (where this mode is dominant), and the needs of the dominant capitalist mode of production (where this is dominant). So, the paradox is that the preservation of the non-capitalist mode is actually in the interests of the capitalist mode, at least when the capitalist mode is dominant. As Taylor (1979: 227) states:

> This articulation of one practice within another is governed both by the reproductive requirements of the capitalist mode, and by the restrictions placed on this articulation either by the limits within which the penetrated instance can operate, as set by the non-capitalist mode of production, or by the continuing reproduction of elements of the non-capitalist mode.

Finally, and related to the previous point, this theory contends that the mode of production and the articulation between different modes of production determine the "phenomenal appearances" of the social formation. Once again Taylor (1979: 274) makes this clear:

> economic features [such] as urban unemployment, combination of different types of labour, and accumulation of indigenous capital . . . [are] analysed as forms whose determinants – the changing reproductive requirements of the industrial capitalist mode and the level of resistance of the non-capitalist mode or division of labour – were necessarily absent in their phenomenal appearance.

A critique

Functionalism

Modes of production theory has two potential strengths compared with modernization and underdevelopment theories. First, it avoids (at least to some extent) the problem of conceptualizing a supposedly "normal" and ahistorical path of capitalist development. Secondly, it potentially leads to a politics that recognizes subordination within the international political economy and exploitation within the nation-state. In other words, modes of production theory represents the most sustained attempt to transcend the "internal–external" dichotomy that plagued the modernization–dependence debate. The result is, or at least could be, "a politics sensitive to local opportunities and struggles". This is preferable to the logic of Frank's and Wallerstein's positions, which "deny the very possibility of meaningful political action in Third World countries" (Corbridge 1986: 61).

Nevertheless, these would-be strengths are never completely realized because modes of production theory is weakened by a functionalist (and, it is argued below, evolutionist) methodological approach. This is because the theory explains social change as a product of a *necessary logic of capitalism*. As the quotations from Taylor, Wolpe and Bradby above make clear, the continued existence of non-capitalist modes is explained by their necessary function of cheapening the reproduction costs of the dominant capitalist mode. But this begs the question of why non-capitalist modes broke down in any society. This leads to the inconsistent and circular reasoning that characterizes all functionalist methodologies. Corbridge's (1986: 63) basic criticism of this approach addresses the problem:

If the PCMP [pre-capitalist mode of production) survives (as in the Bantustans) then that is evidence of its functionality for capitalism; and if it does not . . . then that too is evidence of capitalism's functional requirements. (See also Booth 1985: 774–6)

The implication of this theory is that *any* social change can be explained as the product of the "reproductive requirements" of capitalism. This tendency to "read off" changes in the periphery in terms of the requirements of capital is not, of course, exclusive to this theory. The previous chapter showed that mainstream development theories also explained social change in terms of the needs or actions of the metropolis, so that modernization or underdevelopment were ultimately caused by the Western world.

The roots of this functionalist methodology can be traced back to the original Enlightenment project, but it was the development of Althusserian Marxism in the 1960s and 1970s that had the more immediate impact on development sociology (see for example Hindess & Hirst 1975, Taylor 1979, Meillassoux 1981, Peet 1991, 1992). A re-examination of this methodology (see also Chapter 2) is important, and not only for the obvious reason that it has directly influenced modes of production theory. Perhaps more importantly, though also more controversial, an examination of the problems of Althusserian Marxism is useful because its methodological errors are indicative of the problems of development theory *in general*, particularly radical development theory. Underdevelopment, world systems and dependency theory share similar methodological weaknesses, even if they do not explicitly acknowledge the Althusserian influence. So, an examination of this methodology is useful, as a critique both of modes of production theory and of other radical approaches. In a sense then, in discussing the methodology of Althusserian Marxism, the methodology of *all* radical development theory is being put under scrutiny. This discussion in turn leads back to a discussion of the root causes of the impasse in the sociology of development.

Althusserianism and the impasse in the sociology of development

The work of Louis Althusser and his followers should be seen in the context of debates in the French Communist Party and its attempts to move beyond Stalinist orthodoxies (for a commentary see Benton

1986: ch.1). Althusser's main concern was to break with the "economism" of orthodox Marxism and therefore reject base–superstructure and stagist models (see Chapter 2). More relevant to this discussion however is that he was also equally hostile to empiricist methodologies, which simply took the "facts" as given. For instance he wrote (Althusser & Baliber 1979: 105) that:

> we must once again purify our concept of the theory of history, and purify it radically, of any contamination by the obviousness of empirical history, since we know that this "empirical history" is merely the bare face of the empiricist ideology of history.

His former British followers, Barry Hindess and Paul Hirst (1975: 312), are even more explicit:

> Marxism, as a theoretical and a political practice, gains nothing from its association with historical writing and historical research. The study of history is not only scientifically but also politically valueless.

The "empiricism" of history is replaced by the construction of *concepts*, which "are formed and have their existence within knowledge" (ibid.: 1). Theories and concepts are therefore *a priori* constructs, which are derived from "theoretical practice" (for an extended definition, see the glossary in Althusser & Balibar 1979: 316) that is completely separated from the empirical world. The concept that is given determining status is the mode of production, which is defined as "an articulated combination of relations and forces of production structured by the dominance of the relations of production" (Hindess & Hirst 1975: 9, see also Balibar in Althusser & Balibar 1979: part III). It is this concept that explains social change. Political action may have a "relative autonomy" from the (economic) mode of production, but is always "in the last instance" explained by it.

It is precisely this methodology that articulation theorists share with other theories in the sociology of development. Changes in the Third World are "read off" from the reproductive requirements, or needs, of an *a priori* conception of (global) capitalism. This conception is used to explain the preservation of non-capitalist modes (as in Meillassoux 1981), the destruction of non-capitalist modes (Warren

1980), the development of the periphery (Warren 1980, Rostow 1971), or the underdevelopment of the periphery (Frank 1969a,b, Wallerstein 1974). It also subordinates agency to structure, and assumes that social phenomena are explained by their functionality for capitalism, rather than by the actions and struggles of human beings themselves. Added to this is a spatial dimension, in which the structure is determined by the needs of *Western* capitalism or the capitalist mode within a peripheral social formation. This approach therefore rules out, on *a priori* grounds, any meaningful political action by social actors within the Third World, because they are, by definition, passive actors, directed by the whims of the capitalist mode of production.

Therefore, the "articulation" of modes of production should not be assumed away as part of a "logic of capital", but should be grounded in a historical analysis of social struggles, in a particular location, at a particular time. To return to the example of South Africa, Michael Burawoy (1976; see also Cohen 1987: 105–6) has argued that the cheapening of labour costs under segregation and apartheid was far more contradictory than Wolpe implies. He asks the following questions: in what respects is labour cheap, and who benefits from it? Certainly it was cheap for South African mine-owners, but the costs of policing the system of migrant labour were actually very expensive for the South African state. Moreover, the process of labour recruitment to the emerging capitalist mode of production in late nineteenth-century Transvaal and twentieth-century South Africa was far more problematic than modes of production theory allows. This is because people are not as passive as the theory assumes, and there was widespread (and ingenious) resistance to labour recruitment by the mine-owners (see for example Delius 1980).

Similarly, "economic" relations cannot be given theoretical priority over "non-economic" ones. Class (which, I stress below, is not a purely economic relation) is a very important concept in Marx's thought, but this does not mean that other social factors are reduced to "it". Any suggestion that this is the case implies that class is a "thing" and not a social relation, and it therefore leads back to the economism that post-Marxists so rightly criticize. To take one example, Meillassoux's (1981) attempt to explain the oppression of women as a "mode of reproducing" the labour force should be rejected because it does not explain why it is the man who works in the labour market and the woman in the subsistence sector (see Mackintosh 1977, Bozzoli 1983). Such a functionalist "logic of capital" approach has an economistic starting point, which is separated from the real world and which therefore attempts to explain the oppression of women in terms of its *concept* of capitalism.

Structuralism therefore replicates the errors of the theories discussed in the previous chapter. It takes for granted that which is supposed to be proved, and assumes that capitalism automatically secures its own reproductive requirements. The real history of human societies is far more complex and contradictory than this, as the South African example above illustrates. To recognize the existence of "phenomenal forms" in capitalist society is not the same thing as assuming that these forms automatically serve the needs of capital. As Corbridge (1986: 67) argues,

> there is nothing in the concept of capitalism itself which should lead us to expect that it must have X, Y or Z development (or underdevelopment effects). Such contingencies are not forged at this macro-theoretical scale. The reproduction of capitalist relations of production clearly presupposes the existence of definite conditions of existence (private property relations, for example, and free wage labour), but it tells us nothing about whether they will be secured, or about the all-important forms in which they are made flesh. (See also Mouzelis 1980)

Contrary to the arguments of many Marxists, both Althusserian and orthodox, Marx (1989: 48, the second emphasis is mine) was equally clear on this issue:

> To present the laws of the bourgeois economy, it is not necessary to write the *real history of the production relations.* But the correct analysis and deduction of these relations as relations which have themselves arisen historically, always leads to primary equations . . . which point to a past lying behind this system. These indications, together with the correct grasp of the present, then also offer the key to the understanding of the past – *a work in its own right.*

At this point we arrive back at the extended discussion of Marx that I undertook in Chapter 2. It should be clear that, in separating structure and struggle and in proposing a view of capitalism in which its needs are automatically met, structuralist Marxism is guilty of the fetishism of social forms that Marx so convincingly criticized. This fetishism and functionalism are rooted in modes of production theory's account of the relationship between theory-building and empirical research. So far this point has been only implicit, but in the next section I hope to make my point more clearly. This section examines the relation between capitalism and unfree labour, and discusses the state, and in so doing leads us back to a discussion of the wider relation between theory and empirical research, and structure and agency. This in turn provides a convenient route back to a further discussion of the impasse in the sociology of development.

Capitalism, unfree labour and the state: the relationship between theory and method, and the impasse in development theory

Capitalism, unfree labour and the state

As I showed in Chapter 2, Marx argued that in capitalist society the wage-labourer is "free" in two senses. On the one hand, she is free to sell her labour power as a commodity to any potential employer – that is, the worker in capitalist society differs from the peasant in feudal society in that the former has a choice of to whom she will sell her labour power. On the other hand, freed or separated from direct access to the means of production, the worker is compelled to work for a wage in order to obtain the means of subsistence (food, clothing and shelter). So, to over-simplify, the peasant is compelled to work for the landlord through laws, customs and statutes. In capitalist society, the worker is compelled to work for the capitalist, not for legal or political reasons, but because of "the dull compulsion of economic relations" (Marx 1989: 85).

However, as the example of South Africa above makes clear, capitalism does coexist with unfree labour. Moreover, there are many examples that reinforce this observation. Unfree labour continues to exist in many parts of the Third World, and indeed the state in metropolitan countries has regulated the capital–labour relation – the guest worker system in Germany and the Youth Training Scheme in Britain immediately spring to mind.

Of course one explanation for this persistence, at least in the case of the periphery, is the notion of articulation, but, as I argued above, there are problems with this conception. Others imply that Marx was simply wrong (see Cohen 1987). Still others (for the best exposition, see Miles 1987: 196–222), influenced by the articulation school, argue that capitalism and unfree labour are compatible but that unfree labour is an anomaly (albeit a necessary one). The problem with this argument is that it falls between the two stools of evolutionism and functionalism. Unfree labour is an anomaly because capitalism is based on free wage labour (evolutionism), but it is a necessity because it is based on the needs of capital at a particular time (functionalism).

Once again, the problem with this approach is that it conflates theoretical construction or the elaboration of models with the empirical world, and tries to impose the former on the latter. Marx's analysis of the historical origins of capitalism is not a universal model or a general theory of the capitalist mode of production, as I showed in Chapter 2. Instead, it is an account of how certain social forms, such as the commodity, appear as "natural" but are in fact the product of historically constituted social relations (Williams 1978b). Therefore, just because "free" labour was the "norm" in England (and even this assertion is questionable – see Corrigan 1977), it does not follow that all "capitalisms" must follow the English path. So, although Miles is at pains to separate his analysis from a unilinear theory of history, his assertion that unfree labour is an anomaly leads him back to precisely the trap of regarding England as the norm and all others as anomalies (see Brass 1988: 185). Indeed, this evolutionism is repeated by articulation theory, which argues that "normal" capitalism (of an English type) is different in the periphery, and so it therefore must have "articulated" with non-capitalist modes. In the real world, all are in a sense the norms and all are anomalies. The forms of labour characteristic of particular capitalisms are likely to, and indeed do, vary. Marx (1989: 27–8) argued along these lines when he stated that:

> The specific economic form, in which unpaid surplus labour is pumped out of direct producers, determines the relationship of rulers and ruled, as it grows directly out of production itself and, in turn, reacts upon it as a determining element . . . This does not prevent the same economic basis – the same from the standpoint of its

main conditions – due to innumerable different empirical circumstances, natural environment, racial relations, external historical
influences, etc., from showing infinite variations and gradations in
appearance, which can be ascertained only by analysis of the empirically given circumstances.

This leads on to a second, and equally important, point, which
takes us back to the heart of the impasse. The arguments that capitalism is based on "purely economic" forms of labour control, and thus
that unfree labour is an anomaly, falsely separate different levels in a
social formation. The state is relegated to a derivative, political level,
and class relations are located purely at the determining economic
level (see Poulantzas 1973: 191–4). Such an approach downplays the
prospect of any systematic analysis of the state, because it is said to be
derived (albeit in the last instance) from the economic level. This is of
direct relevance to any discussion of the impasse because the state in
peripheral (and, I would argue, all) capitalist societies cannot be as-
signed a derivative or superstructural role – which, critics of the impasse have argued, is one example of Marx's "economic determinism" (Mouzelis 1988: 35–40, ibid. 1990). The state should not be
derived from a determinant "economic" sphere, but should instead
be conceptualized on the basis of the historical emergence of capitalist social relations. As Corrigan and Sayer (1981: 27) state:

> Marx's argument is not that law and politics are separate from
> production and its relations as such, but rather that in commodity
> production the social relations of production *themselves* take the
> forms of apparently exclusive political and economic spheres.
> "Polity" and "economy" are in other words but different facets of
> one and the same set of production relations.

In other words – and this has great implications for "post-Marxism"
(and neoliberalism, which is discussed in Chapter 6) – a separate
"economy" or "political system" are forms of existence of capitalist
social relations. Marx was clear on this and argued that the state was
not part of a "superstructure" in capitalist society. For example, he
documented the role of the state in the "primitive accumulation" of
capital in England, and argued that "force" (a supposedly "political"
relation) "is itself an economic power" (Marx 1976b: 916).

"Back to the impasse"

The discussion of this particular debate, namely the relationship between capitalism, unfree labour and the state, once again leads us back to the impasse. Contrary to the claims of structuralist Marxism, the political state does not simply serve the needs of capital, nor is it part of a so-called superstructure. Similarly, unfree labour is not simply an anomaly, but is a constitutive part of "actually existing capitalisms". It is these kinds of reductionist assumptions that have been forcefully challenged by post-Marxist writers in the development field, and so it is now appropriate directly to confront debates around the impasse, and the ways in which Marxism has contributed to it.

The writer who first identified an impasse was David Booth (1985). The impasse was seen as "the result of a generalized theoretical disorientation affecting in different degrees all of the main positions in the radical development debate" (ibid.: 776). Underdevelopment and related theories are rejected on similar grounds to those outlined in the previous chapter, but what is more important here is his similar rejection of Marxism. His main contention is that Marxism has a "metatheoretical commitment to demonstrating that what happens in societies in the era of capitalism is not only explicable, but also in some stronger sense necessary" (ibid.: 773). So, in the two examples that I have discussed in this chapter, the preservation of non-capitalist modes of production and unfree labour relations are "explained" simply by the "needs of capital". Empirical research is subordinated to this *a priori* theoretical construction (see also van der Geest & Buttel 1988, Sklair 1988). For development sociology to move beyond this stagnation it "must be freed . . . from Marxism's ulterior interest in proving that within given limits the world *has* to be the way it is" (Booth 1985: 777).

As an *explanation* for the impasse, Booth's paper is a powerful one, not least because it shouts what others have only whispered. What is less clear are his suggestions for a *way out* of the impasse. For example, he argues that some Marxist concepts are still useful (ibid.: 777), but does not specify which should be dispensed with and which should be retained. It has been left to others to make explicit suggestions of ways of transcending the impasse. A common argument is that development sociology should focus on empirical and comparative research and construct "middle-range" theories out of this work

(Mouzelis 1988, Evans & Stephens 1988, van der Geest & Buttel 1988, Sklair 1988). I share these views and expand on them below (and register my concern about how this might be carried out), and in Part II of this book.

But, for the moment, I want to concentrate on the suggestions of one writer who has developed Booth's arguments and suggested some ways of moving out of the impasse – Stuart Corbridge. He argues that the way out of the impasse is the development of a "post-Marxism" that rejects the functionalism and determinism of modes of production and other development theories (1990: 628). This is most clear in his discussion of the work of Hindess and Hirst, who broke with the "high Althusserianism" of their early work, discussed above, and laid the framework for a "post-Marxist" alternative (see Corbridge 1986: 66–7, ibid. 1990: 629, Hindess & Hirst 1977, Cutler et al. 1977, 1978). In their post-Marxist work, the conceptual framework of Marxism is said to be tautological and economically reductionist because it is assumed that the economy can secure its own "non-economic" conditions of existence. To return to the three examples in this chapter, this would include the preservation of non-capitalist modes, of unfree labour, and of a state that served the interests of capital. As Hirst (1977: 131) has argued,

Either economism, or non-correspondence of political forces – that is the choice which faces Marxism.

Similarly, Hindess (1978: 96–7) has argued that,

The choice for Marxism is clear. Either we effectively reduce political and ideological phenomena to class interests determined elsewhere (basically in the economy) . . . Or we must face up to the real autonomy of political and ideological phenomena.

What is clear from these two quotations is that Hindess and Hirst give only two choices – *either* general determinism (by the concept of the economic mode of production), *or* absolute autonomy from this determinism. So, to return to one example above, the state is either determined by the economy or autonomous from it. The first proposition is wrong because it is based on a theory in which the determining role of the economy is built into the discourse itself (Corbridge

1990: 629). Therefore, according to the logic of the theory, the second proposition must be correct. Although the move away from theoreticism and structural-functionalism is welcome, there is a great danger in this "either/or" approach. For, despite Corbridge's protestations (1990: 628–9, 1986: 66–7), this dualism has led Hindess and Hirst to move away from a *theoreticism* divorced from the empirical world to an *empiricism* that takes for granted, and therefore naturalizes, historically constructed social forms. Hindess and Hirst's post-Marxism is therefore as guilty of fetishism as the structuralist theory it leaves behind. This is hardly surprising given that the two views, based on a rigid dichotomy between the theoretical and empirical world, are in effect a "mirror image" of each other. As Meiksins Wood (1990: 128, see also Meiksins Wood 1986) points out,

> It soon turned out that Althusserianism had simply replaced – or supplemented – the old false alternatives with new ones. Marxists had in effect been offered a choice between structure and history, absolute determinism and irreducible contingency, pure theory and unalloyed empiricism. It is not surprising, therefore, that the purest theoreticists of the Althusserian school became the most unalloyed empiricists of the post-Althusserian generation.

What this discussion suggests is that functionalism, economism and theoreticism should all be rejected as Booth, Corbridge and others suggest. However, they should not be replaced by an empiricism that naturalizes social forms or autonomizes social practices. The notion that "the economic determines the political" is replaced in Hindess and Hirst's (and their collaborators') auto-critique by the notion that "the economic does not determine the political" (see, among others, Cutler et al. 1977: 222, 224, 226, 227). Booth and Corbridge both correctly reject the economism of the first premise, but say nothing about the second. However, as the discussion of the state above made clear, it is the *common starting point* – that is, the separation of discrete, separate "levels" – of *both* contentions that is inadequate. The implications of this argument for social analysis are great, because it suggests that factors such as "class", "gender" or "state" cannot be segregated to their own particular "sphere" or "level". Class is thereby *not* a purely "economic" relation, just as the state is not a purely "political" relation. By separating *social phenomena* into compartmentalized units, Hindess and Hirst (marks I and II), automatically take "things as they are" and so reproduce the "phenomenal forms" of social relations. That is, they are guilty of *fetishizing* capitalist social relations.

This is not to suggest that the new "post-Marxists" in development theory have followed this road, but there is a danger that they may replicate the fetishized categories of Hindess, Hirst and others. To take another example, Booth is rightly scathing of grand theories of development but he takes a rather eccentric view of one work that is similarly critical. This is Robert Brenner's (1977) justly famous article on the origins of capitalist development. In this work Brenner explains the dynamic nature of capitalism in the "core" as a product of the historical emergence of a capital–free wage labour relation and of the generalized commodity production and "free market" that this gives rise to. Booth (1985: 770) rightly describes this essay as brilliant, but also complains that it "does not give us what many people have looked for in the mode of production literature, namely, a genuine third position in the debate over colonial and contemporary development in the Third World". I think that Booth is undoubtedly correct, but that is one of the great *strengths* of Brenner's work – it is not a search for a "new position" because such a position, forged at the level of "high theory", will automatically subsume practice to theory and agency to structure. Booth (1985: 770) assumes that Brenner shares Warren's optimism concerning capitalism in the post-colonial period, but the whole thrust of Brenner's paper is a challenge to the "developmental" logic that assumes that the effects of capitalism can be examined independently of the actions of human beings. In other words, Brenner's work reconstitutes the unity of structure and agency that development and modes of production theory separate. As Ruccio & Simon (1986: 216) state, for the Brenner approach "it is not a matter of a mechanistic, deterministic process" (see also Foster-Carter 1978: 77, for a "Brenneresque" feminist critique of modes of production theory, see Bozzoli 1983).

Booth's ambiguity about theory/practice and structure/agency is also apparent in his conclusion when he recommends the preservation of some lower-order concepts derived from Marx, but also suggests that there is a case for "a relative (and hopefully temporary) shift of emphasis in the sociological development debate from theory to metatheory" (1985: 777). The problem with this proposal is that it proposes no way out of the structure/agency dichotomy, and is likely to replace one "grand theory" with another.

Finally, Booth and Corbridge both caricature Marx's work and so, once again, there is a distinction to be made between Marx's method and the method of his followers. Althusserian Marxism, and therefore modes of production theory, begin with the *concept* of the mode of production, which determines non-economic institutions and the social practice of human beings. Marx, on the other hand, did not begin with *a priori* dogmas. Instead, Marx & Engels (1982: 42, my emphasis) stated that:

> The premises from which we begin are not arbitrary ones, not dogmas, but real premises from which abstraction can only be made in the imagination. They are the real individuals, their activity and the material conditions under which they live, both those which they find already existing and those produced by their activity. These premises can be verified in a *purely empirical way*.

This point is consistently made *throughout* Marx's work, and is not part of Marx's humanist youth, as Althusserians contend. For example, Marx (cited in Sayer 1979: 92) explained his methodology in *Capital* in the following way:

> I do not start from "concepts" . . . What I start from is the simplest form in which the labour product is represented in contemporary society, and this is the "commodity". I analyze this, and indeed, first in the form in which it appears . . . Thus it is not I who divide "value" into use-value and exchange-value as oppositions into which the abstraction "value" divides itself, but the concrete social form of the labour product.

This famous – but strangely neglected – passage from Marx, describing his methodological approach, leads us nicely back to the questions of articulation theory and unfree labour. The fact that different capitalisms utilize different forms of labour control does not mean that the concept of the capitalist mode of production has to be abandoned. Once again, structuralism presents us with a false dichotomy – either a capitalism based on one form of labour control (wage labour), or the abandonment of the concept of capitalism (as a social form) itself. In other words, the choice is between a "pure" capitalism and purely contingent methods of labour control. But there is a third choice, which is based on *different capitalisms*, based on differing methods of labour control, whose commonalities and differences are explained through comparative and historical analysis.

Marx's critique of political economy is thus not a grand theory of a pure capitalism but a "relational analysis. It allows us to ascertain the conditions of existence of particular kinds of experiences, but it tells us nothing at all about how these conditions originally came about. It is not, in other words, a causal analysis. It merely reveals internal relations: those 'conditions and relations' of particular modes of production" (Sayer 1975: 789, see also Sayer 1987, and Ollman 1976).

Conclusion: Sociology, development and the impasse

Modes of production theory constituted the most sustained attempt to break with the evolutionism of development theory after 1945. It rightly pointed to the existence of different forms of labour control in "actually existing capitalisms", and provided a convincing explanation for the reality of uneven development and the different forms this had taken in various parts of the world. Moreover, in placing some emphasis on the internal structure of peripheral social formations, the theory also potentially gave some considerable weight to the salience of local struggles. However, this potential was never fulfilled because the theory repeated the errors of previous development theories, albeit at a lower level of abstraction. The major problem was the theory's functionalism, in which changes in the Third World were "read off" from the needs of the dominant capitalist mode of production. This was not so very different from (Marxist and non-Marxist) modernization and underdevelopment theories: the former explained development as a consequence of "Westernization", while the latter explained underdevelopment as a consequence of the dominance of Western capital. In each of these theories, the struggles of social actors were reduced to the overriding requirement that the structural needs of the system be reproduced.

The result of this excessive structuralism was that empirical research is carried out on the basis of fitting into a preconceived theoretical model. There was actually very little need for empirical research because the results of, and explanations for, changes were built into the theoretical system. This in turn led to a dogmatism that rejected the empirical findings of scholars who lacked the "correct" *a priori* method (see for example Peet 1989, 1991). It is for this reason that the *sociology* of development largely isolated itself from the impact of the counterrevolution in development theory in the 1980s.

The growing awareness of this impasse has led to the suggestion that the discipline should adopt a "post-Marxist" approach, which would utilize some of the non-functionalist work of radical social theory. This might include regulation theory, post-imperialist theory or structuration theory (see Corbridge 1989, 1990, Spybey 1992). Some of these ideas are discussed in more detail in Part II, but it should be clear from the above discussion that there are some grounds for questioning the utility of at least some post-Marxist approaches. This does not mean that one simply "returns to Marx" to show that he had all the answers. This is precisely the kind of ahistorical dogma that I reject. What *is* useful in Marx's work is a historical and materialist method, *grounded in empirical research*, that shows that phenomenal forms are historically and socially created. Marx's concern was therefore to build theory out of empirical research. This approach differed from an empiricism, that took the "facts" as given, precisely because of its recognition that such "things" are neither universal nor natural.

The impasse is therefore best overcome – and here I agree with the general arguments of those who have recognized the stagnation in the field – by a renewed commitment to empirical research, and the abandonment of the search for a grand theory (see Mouzelis 1988, van der Geest & Buttel 1988, Sklair 1988). This does not mean that theory *per se* should be abandoned, for this would lead back to an empiricism that is simply the other side of the theoreticist coin, and in which "things" are "taken as they are". It is in this way that theory can be built *a posteriori*, on the basis of the recognition that the struggles and actions of "social actors" are important, but also that there are structures that impede their action. These structures might even give rise to certain "laws" (see the next chapter for an example), but these are neither universal nor separable from the action of human beings.

This leads to my final point, which concerns the way out of the impasse and the distinctiveness of sociology. Development *sociology* must transcend its theoreticism and give proper weight to the interdisciplinary nature of development studies. I also believe that it is precisely at this point that it may also preserve its specific character and at the same time move beyond its theoretical stagnation. The discipline should draw on the best empirical work in the field, and use this as the basis for the development of a "grounded theory". There is a wide scope for work on the basis of "suggestive contrasts" (Runciman 1983: 168–93): for example, between the successful development trajectories of east Asian newly industrializing countries compared with the less successful examples in Latin America (Evans 1987, Jenkins 1991, and the next chapter), or between state formation in Africa and east Asia (see Sandbrook 1986).

The impasse is therefore best transcended by a method that is historical, empirical and materialist. I hope that it is clear, however, that any work that does not utilize this method cannot simply be dismissed on *a priori* grounds; it must instead be demonstrated by a direct confrontation with, rather than dismissal of, the discourse itself. I therefore stress again the *a posteriori* nature of theory building. In Part II of this book I expand on these arguments through a discussion of some of the recent changes in the global political economy and some of the political implications of these developments.

Part II

The impasse and the contemporary global political economy

5

The impasse and Third World industrialization

One of the most important changes over the past 20 years in the global political economy has been the unevenness of economic performance between different Third World economies. The oil price rises of 1973–4 and 1979 led to the rapid economic growth of oil exporters. I touch on this phenomenon in the context of Third World nationalism in Chapter 7. This chapter is concerned with the other major reason for economic differentiation within the so-called Third World – the rise of the "newly industrializing countries" (NICs).

The success of some economies, particularly the "four tigers" of east Asia (Hong Kong, Singapore, Taiwan and South Korea), has encouraged the restoration of neoclassical theory as the dominant paradigm in development theory, and the decline of underdevelopment and dependency theory. This chapter examines development theory in the light of the rise of the NICs, especially those in east Asia, which have been taken up as a new "model" in development studies. The theories examined are: (i) neoclassical and modernization theory; (ii) orthodox Marxism; (iii) new versions of dependency and world systems theory, and in particular the theory of the new international division of labour; (iv) regulation theory. The first section outlines the basic contentions of these theories, while the second section provides a critique, and, in returning to the impasse, I suggest that they share a common methodological starting point. In the third section, an alternative approach to the global economy is suggested. Finally, I discuss the east Asian experience in the light of its specific domestic social and political history, and discuss how this was related to wider international political and economic forces.

Competing accounts of the rise of the NICs

Neoclassical and the "new modernization" theory

It should be clear from Chapter 3 that the neoclassical theory of comparative advantage was challenged by the "structuralist" critique in development economics in the 1950s and 1960s. This was accompanied by an uneasy "alliance" with modernization theory, which was less critical of the international economic order but shared the strategic goal of development through industrialization. However, this import-substitution industrialization (ISI) was often led by the state and its main focus was the domestic market of peripheral economies. With the onset of the debt crisis in 1982, it was clear that import substitution had come up against definite limits, and that countries that followed this strategy were still importing more than they were exporting. It was of course on this basis that dependency theory had earlier emerged, arguing that this process reflected the subordinate position of the Third World in the international economic order and the contradictions between the global strategies of transnational corporations and the national strategies of Third World states. But the 1980s presented this theory with a new problem: while the ISI regimes in Latin America went into a massive recession, the export-oriented regimes in east Asia either continued to grow or at least quickly recovered from recession. Neoliberal economists argued that this success was based on the "good government" policy of adopting neoclassical principles. According to Ian Little (cited in Lal 1983: 45, see also Little 1981), "success is almost entirely due to good policies and the ability of the people – scarcely at all to favourable circumstances or a good start".

The implication of this statement should be clear. East Asia is not a "special case", and if all Third World regimes adopt similar policies then they too will enjoy high rates of growth. It is in this way that neoclassical theory has become a new version of modernization theory – the Western model of 1950s' modernization theory has been replaced by the east Asian model of 1980s' neoclassical theory. Influential right-wing social scientists have also recognized the east Asian success stories and regarded them as a capitalist model for other countries to follow (see Berger 1985: 28–30, Fukuyama 1992: 100–108).

According to neoclassical theory, the success of the east Asian NICs can be attributed to three closely related policies. These are: (i) limited government intervention in the economy; (ii) a low level of price distortion in the economy; and (iii) an "outward-oriented" export-promotion strategy (see Balassa et al. 1986). Limited government entails policies such as the removal of state subsidies to industry, the removal of minimum wage legislation and the removal of price controls. These in turn, together with a realistic exchange rate, lead to a low level of price distortion, which facilitates "correct price signals", to which investors and consumers respond. Therefore, a close correlation exists between a low level of price distortion and a high level of economic growth (see World Bank 1983: 60–63). Finally, export promotion allows a government to import goods in which it does not possess a comparative advantage, in return for the export of those goods that it does produce cheaply and efficiently.

This scenario is contrasted with the inherent inefficiency of the import-substitution model of industrialization (Balassa et al. 1986: 19). Although neoliberals recognize that a great deal of Third World industrialization has taken place without export promotion, it is argued that this is inefficient and unsustainable. State protection to domestic industry leads to the subsidizing of expensive producers, and so consumers have to pay the costs in the market place. This problem is intensified by state allocation of licences for selected imports and state regulation of the financial markets. Both these forms of rationing encourage the promotion of speculative, rather than productive, economic activity, as would-be entrepreneurs compete for lucrative import licences or access to state credit (for example, see World Bank 1985: 37–41). It is for these reasons that the ISI model is said to have failed, in contrast to the east Asian model, which promoted exports on the basis of the region's *comparative advantage* in cheap labour.

The east Asian model, or, more precisely, the neoclassical interpretation of it, informs the thinking of International Monetary Fund austerity programmes. The assumption is that east Asian economies have continued to grow because their governments provided the correct policies in which market forces could develop and in which their economies could therefore exercise their comparative advantage. If other countries follow the policies of limited government, export orientation and "getting the prices right", they will attract substantial domestic and foreign investment, and then they too can join the same economic league as the newly industrializing countries. This school of thought therefore argues that the theory of comparative advantage was right all along, and this is the key to understanding successful (and sustained) Third World industrialization.

Orthodox Marxism

I have already demonstrated the developmental logic of orthodox Marxism, and so it should be no surprise that Warren and others (Schiffer 1981) argue that the old colonial division of the world between Third World primary producers and First World manufacturers has been replaced by a new order in which the Third World has rapidly industrialized. The onset of independence gave Third World countries considerable room for manoeuvre in the international system, which has undermined the dominance of a few advanced capitalist countries (Warren 1973: 10–16). Moreover, direct foreign investment by transnational corporations in the Third World benefits the receiving country. This is the case even when there may be substantial repatriation of profits back to the host country, for the simple reason that there is at least an initial inflow of capital, which would otherwise not exist in the receiving country (see Warren 1980: 175ff.). This leads Warren (1980: 176) to conclude that:

> private foreign investment in the LDCs is economically beneficial irrespective of measures of government control . . . To the extent that political independence is real, private foreign investment must normally be regarded not as a cause of dependence but rather as a means of fortification and diversification of the economies of the host countries. It thereby reduces "dependence", in the long run.

It is clear, then, that Warren and others argue that the industrialization of the Third World is a product both of political independence and of investment by transnational companies, which progressively develop the productive forces in the periphery. I criticize this argument below, but one observation can be made immediately, which is the similarity between neoclassical and orthodox Marxist views. Both views are principally concerned with economic growth (on this point, see Seers 1979: 3–8), and both views regard TNC-led industrialization as a major way of achieving this goal.

Dependency and world systems theory revisited: the "new international division of labour"

As already stated, the rise of the newly industrializing countries undermined many of the contentions of at least some versions of dependency theory. However, some of the leading theorists of this school, such as Frank and Wallerstein, began to offer new explanations for the rise of the NICs, and others offered new accounts that owed much to the original formulations of the dependency school. The most famous of these is the theory of the "new international division of labour" (NIDL), which is chiefly associated with the work of Frobel, Heinrichs & Kreye (1977, 1980).

This school of thought argues that the rise of the newly industrializing countries has led to the growth of a new international division of labour. However, contrary to the contentions of neoliberals and orthodox Marxists, this theory argues that the NIDL has sustained, and indeed intensified, dependency in the world system. This is because the NIDL reflects the interests of the Western world as much as the old international division of labour, and because the industrialization of the periphery is not "genuine" and remains dependent on the advanced capitalist countries.

The origins of the NIDL go back to the late 1960s and early 1970s, when the advanced capitalist countries faced a "profit squeeze". High wages won by the working classes of the core countries cut into profits, and so an increasingly international capital began to relocate its activities from core to periphery. In so doing, it took advantage of cheap labour in the Third World, a process supported by the fragmentation of job tasks and hence deskilling, and developments in global transport and communications (Frobel et al. 1977: 80–83).

The new international division of labour therefore represents the increased industrialization of the Third World and the rise of the "post-industrial" society in the core countries (Peet 1986: 81). However, this has not substantially altered power relations in the international system, because the rise of the NICs reflects the needs of capital at the centre. Workers in the Third World produce goods at low prices, which are then exported back to the core countries, where they are consumed at a low cost. The "post-industrial society" is therefore based on the exploitation of Third World manufacturing.

This exploitation occurs through the employment of workers who face long hours, poor working conditions, low pay, a high level of labour intensity, short holidays and rapid rates of labour turnover (Frobel et al. 1980: 350ff.). A high proportion of the workers are women, which in highly gendered societies facilitates the payment of even lower wages and also more rapid labour turnover once women reach an age when they are expected to become housewives (Elson & Pearson 1981). This "super-exploitation" (Frank 1981b: 53–65) is often carried out in export processing zones, where trade unions are banned and there are tax incentives for prospective TNC investors. This is often reinforced by the presence of authoritarian states, and it is this fact that leads Frank (1981a: 188) to argue that:

[t]he imposition of greater exploitation and superexploitation in the Third World as instruments of export promotion and participation in the international division of labour during the world economic crisis must be enforced through political repression. . . . This repression is not accidental or merely ideologically motivated. Rather it is a necessary concomitant of economic exploitation.

So, the NIDL is a product of the changing needs of global capital since the late 1960s. TNCs have relocated from core to periphery to take advantage of a cheap, easily controlled labour force in the periphery, whose products are then exported and sold in the markets of the core countries. This process has not reduced dependency, precisely because it reflects the needs of TNCs in the core, and so cannot be described as "autonomous" capitalist development (Frank 1981a, Hart-Landsberg 1984). As Frobel et al. (1977: 83) state:

the presently observable world-wide industrial relocation of manufacturing . . . is the result of a qualitative change of the conditions for capital expansion and accumulation enforcing a new international division of labour.

Regulation theory

This theory constitutes the most systematic attempt to theorize the changes in the global political economy since the late 1960s (see Aglietta 1987). The regulation school utilizes two key concepts to explain the changes: the *regime of accumulation* and the *mode of regulation*. The regime of accumulation "describes the stabilization over a long period of the allocation of the net product between consumption and accumulation; it implies some correspondence between the transformation of both the conditions of production and the conditions of the reproduction of wage earners" (Lipietz 1986: 19). So a regime of accumulation is based on a particular set of conditions and social practices that promote rapid economic growth. However, these conditions of existence of a regime of accumulation are not automatically secured as "there is no necessity for the whole set of individual capitals and agents to behave according to its structure" (ibid.). What must also exist is "a materialization of the regime of accumulation taking the form of norms, habits, laws, regulating networks and so on that ensure the unity of the process, i.e. the approximate consistency of individual behaviours with the schema of reproduction" (ibid.). This "body of interiorized rules and social processes" is called the mode of regulation (ibid.). In other words, it can be seen as a kind of "support framework" for growth regimes, pulling together firms, banks, retailers, workers and so on into a regulated network (see J. Allen 1992: 186).

The late 1960s saw a crisis of one particular regime of accumulation – Fordism. This was based on the economies of scale that derived from the mass production of standardized products, where high productivity coexisted with relatively high wages, which in turn served to unite production and consumption, thereby avoiding a crisis of underconsumption. However, in the 1960s this regime of accumulation went into crisis as the labour process appeared to reach its technical and social limits (Harvey 1989: 147) and productivity gains slowed down, which created a crisis of profitability (Lipietz 1986: 26). This in turn created the new problem of a potential crisis of underconsumption – too many goods chasing too few buyers – which could be alleviated only by an increase in buying power, which meant wages would have to increase. However, this would also intensify the crisis because higher wages would cut into already declining profitability (ibid.).

Capital's response to this crisis varied, but it can best be described as the adoption of *flexible* employment practices. Indeed, some writers (Harvey 1989: 147) argue that Fordism has been replaced by a new flexible regime of accumulation, which "rests on flexibility with respect to labour processes, labour markets, products, and patterns of consumption". This all-embracing concept of flexibility is therefore used to describe the advent of computer-controlled manufacturing systems, which lead to a reduction in the demand for labour, and a shift in the character of the core labour force from machine operators to skilled technicians. This flexibility at the work-place is reinforced by flexible work-groups, whose all-round skills help to facilitate a shift from the production of mass goods to small-batch, short-life products.

This shift from mass to flexible production is also reflected within the international economy, as transnational companies shift certain parts of their production to the periphery. This relocation is based on the search for markets and for cheap labour (Lipietz 1986, 1987). The result of this relocation has been the rise of "bloody Tayloriz-ation" and "peripheral Fordism" in the Third World (Lipietz 1986: 31). The former is "a case of the delocalization of precise and limited segments of sectors in social formations with very strong rates of ex-ploitation (in wages, duration and intensity of labour, etc.), the prod-ucts being mainly re-exported to the core" (Lipietz 1986: 31). The latter is an authentic Fordism "based on the coupling of intensive ac-cumulation and the growth of markets. But it remains *peripheral* in the sense that in the global circuits of productive sectors, qualified employment positions (above all in engineering) remain largely exter-nal to these countries" (ibid.: 32). What Lipietz appears to be argu-ing, then, is that the growth of peripheral industrialization is the product of the relocation of TNCs to the periphery, which has created the "bloody Taylorism" of export processing zones, or is the product of the growth of Fordist practices in selected countries.

The similarities between this perspective and the theory of the new international division of labour should be clear, and I return to this point in my critique below, but first to summarize: regulation theory periodizes capitalism in terms of modes of regulation and regimes of accumulation, and its advocates argue that, since the late 1960s, the Fordist regime of accumulation has broken down. This has been re-placed by more flexible, "neo-Fordist" patterns of accumulation, which include the growth of labour-intensive and even Fordist manufacturing in the Third World.

A critique

The theories discussed so far are obviously divided on many key issues, and most writers would argue that they are ultimately incompatible. Nevertheless, although recognizing the major gulf that separates these theories, in this section I want to argue that they do in fact share certain common assumptions, which are rooted in their similar methodologies. Therefore I criticize these theories in general terms, by examining the evidence, and then move on to more specific criticisms of each particular theory.

General comments

Although the terminology of the theories might differ, the theories actually share surprisingly similar assumptions. All are concerned with the relocation of First World manufacturing to the Third World, and all explain the rise of the NICs in these terms. The terminology might differ, but the explanation is very similar – TNCs relocated to the Third World: (i) because of the latter's "comparative advantage" in cheap labour (neoclassical theory); (ii) because of the exploitation of cheap labour in the periphery (NIDL and regulation theory). Warren et al. are less concerned with explaining the rise of the NICs, but in so far as they have an explanation, it is probably closer to the first approach (on this point, see Seers 1979). This similarity has also been noted by Amsden (1990: 19), who writes that:

> [f]or neo-classical economists, economic development on the basis of such exports [of labour-intensive goods] is taken as a confirmation of the theory of comparative advantage. For global Fordists, economic development on the basis of such exports is attributed to capitalist "crisis" and global restructuring. The terminology is different, but the emphasis on a new international division of labour as an explanation of economic development is quite similar.

It is undoubtedly true that some TNCs did relocate to some peripheral countries from the late 1960s (or even earlier). The problem with the theories discussed is that they take one tendency (that is, the relocation of some labour-intensive manufacturing from First to Third World) and construct their own particular "grand theory" from this one particular process. This common approach is not entirely surprising because they all share a common starting point, which is that they all "explain", or "read off", changes in the periphery from the changing needs of the centre. In other words, changes in the periphery are primarily explained by changes in the core. We therefore arrive back at the impasse, and the methodological assumptions associated with it.

This common methodological weakness leaves all of the theories facing difficult empirical questions. First, the proportion of global direct foreign investment in manufacturing going to the periphery actually *fell* between 1960 and the early 1980s (Jenkins 1987: 13), and so the proportion of foreign investment in manufacturing flowing between First World nations *increased* in this time. This period was supposedly *the* period of the creation of a new international division of labour. Secondly, the share of manufacturing production in the Third World (including the NICs) in 1984 was 13.9 per cent of total global manufacturing. This represents an increase from 12.2 per cent in 1966, but it is actually *less* than the 1948 figure of 14 per cent (Gordon 1988: 33). Thirdly, export processing zones (EPZs) are not as significant as either the right or the left assumes. The right regards them as the providers of the institutional framework for the realization of the periphery's comparative advantage in cheap labour, while the left sees them as examples of capitalist "superexploitation". However, in South Korea, EPZs accounted for less than 5 per cent of Korea's total exports in the early 1980s (Edwards 1992: 110); employment in EPZs rarely accounts for more than 5 per cent of total industrial employment; and evidence from India, the Philippines and Taiwan indicates that in each case less than 10 per cent of all manufactured exports came from EPZs (Jenkins 1987: 132). Moreover, many companies in EPZs are not TNCs but are in fact local companies (ibid.: 133). Fourthly, the importance of direct foreign investment varied between different newly industrializing countries, and in some the amount was not very high. It was a key factor in Singapore, but in Taiwan – a country that was among the most successful in attracting foreign investment – foreign firms accounted for only 5.5 per cent of capital formation between 1962 and 1975 (Wade 1990: 149). In South Korea direct foreign investment was even less significant, as I show below. Once again, these data suggest that the industrialization of the periphery cannot be "read off" from the globalizing activities of core companies.

I expand on this point below, as I examine the problems of each particular theory.

Neoclassical theory: a critique

The major criticism that can be made of neoclassical theory is that it almost totally neglects the role of the state in the economic development of east Asia. This is unsurprising, given the theory's key assumption that unregulated markets lead to equilibrium and balanced or "even" development. A number of writers influenced by neoclassical theory recognize that states in east Asia did intervene in the economy (see Balassa & Williamson 1987: 5), but they argue that the success of the Asian NICs "has been achieved *despite* intervention" (Lal 1983: 46). I criticize this contention below through an examination of the evidence, but it should be clear that even this acceptance of state intervention flies in the face of Deepak Lal's confident claim that the success of the NICs undermines state-centred approaches to development (ibid.: 17–18).

The claim that growth occurred despite intervention seems all the more problematic when one compares state activity in Latin America with that in east Asia. In 1986, state spending in Taiwan constituted 25 per cent of GNP, compared with 15 per cent in Colombia (1987) and 25 per cent in Uruguay (Jenkins 1990: 48). Moreover, the share of public enterprise in total fixed investment was higher in east Asia in the 1970s than it was for most Latin American nations (ibid.). In Taiwan, the state accounted for 57 per cent of industrial production in 1952 and, although there has been significant privatization since then, the state's share of gross domestic investment still stood at 50 per cent in 1980 (Amsden 1985: 93). In terms of foreign investment, if anything the state in Taiwan and South Korea has been more restrictive than the state in most Latin American countries (Jenkins 1990: 49). Although Singapore operates a far more liberal foreign investment regime, the state can hardly be described as "laissez-faire". The state directly owns many strategic industries, and the state-run insurance fund generates such enormous revenue from high taxation that it is used as a major instrument of public finance for savings and investment (Harris 1986: 62). Even Hong Kong, which is perhaps closer to the claims of the free marketeers (though, given its size, location and urban concentration, hardly a model for others to follow), has close informal ties between government and business (Deyo 1987b), massive state ownership of land, which is "sold" on lease-hold as a means of generating revenue, and heavy state intervention in the housing market (Drakakis-Smith 1992: 167). Such data alone do not prove that state activity is efficient, but they certainly demon-strate that it is *not necessarily inefficient*. The neoclassical case is thus severely weakened by these figures alone. Nevertheless, Lal (1983: 46) and others argue that there were certain free trade (rather than laissez-faire) policies that promoted economic growth in east Asia, and it is to these that I now turn.

At the heart of the neoclassical case is the claim that free trade poli-cies were adopted in east Asia. These were export orientation and "getting prices right". However, both of these contentions are ques-tionable. The distortion index used by the World Bank in the *World Development Report* 1983 (see above) is open to doubt; for example, the pro free market regime in Chile was said to have a high rate of price distortion. Moreover, even if one accepts the validity of the fig-ures, they explain only one-third of the differences in growth rates, so there must be other factors not linked to market forces. Finally, the relation between price distortion and growth might not be a causal one, and might in fact be the result of a third factor, such as structural problems that slow down economic growth, which in turn leads to price distortions (on these criticisms, see Jenkins 1992: 193–4, Evans 1993: 57–64). The focus on outward orientation might be similarly misplaced; for example, some moderately outward-oriented coun-tries stagnated between 1973 and 1985, while some inward-oriented regimes enjoyed high growth. The fact that the most strongly inward-oriented regimes are the poorest Third World nations does not itself mean that greater trade *per se* is the answer; instead, this correlation might "simply reflect the greater obstacles to growth faced by the poorest nations" (Jenkins 1992: 194). These obstacles cannot simply be explained as a product of the lack of international trade of a par-ticular economy, but might actually reflect the highly unequal pro-duction structures that exist in the international economy. I expand on this point below.

This point leads us back to a consideration of the role of the state in promoting industrialization in "late", and potentially weak, develop-ing societies. As already stated, neoclassical theory stresses the case for limited government and "market forces", and, although recognizing that the state may have played a role in east Asia's economic development, argues that growth would have occurred anyway (see Lal 1983 above, also Balassa & Williamson 1987: 6–15). However, there are strong grounds for questioning this view and arguing that, without state intervention, export-led growth would not have occurred. Indeed, some writers have accurately recognized that in east Asia the state consciously "got prices wrong" in order to boost export performance (see Amsden 1989: 139–55). For example, in South Korea from the 1960s, many firms exported at a loss, and so were in effect subsidized by the state. This support applied to labour-intensive industries such as textiles, as well as to heavy industry and "high-tech" goods such as iron and steel and computers (Amsden 1989: 143, 316, 82). Once export targets were met, losses could be recovered through sales in the *protected* domestic market. This market was protected by limited access to licensed imports, which were generally awarded to the most successful exporters. In this way, high export performers were rewarded with access to scarce imports, and were therefore able to recover losses made on exports. This practice was not the only form of state intervention, and there were also price controls and state incentives to certain strategic industries, especially heavy industry. Hamilton (1986: 83) states that the average deviation of production prices from actual prices increased from 7.8 per cent in 1966 to 11.8 per cent in 1978. The implications of these policies are quite clear: the export promotion that neoliberals are so quick to support was made possible only by strategic state intervention. If the South Korean state had left the economy to "market forces", then loss-making exports would not have taken place.

A further implication is that the "developmental state" in Taiwan, South Korea et al., and the social forces that created it, are unique products of the political and social history of those particular countries (see Hamilton 1987). It is therefore impossible to construct a "model" on the basis of the success of these economies. As I showed in the first part of this book, this is one of the major weaknesses of development theory, including radical development theory. Instead, all that development theory can do is suggest reasons for the "success stories" in east Asia and the comparative failures elsewhere (for useful attempts, see Evans 1987, Jenkins 1991). It cannot impose an ahistorical and asocial model of one region on another, as the IMF and World Bank try to do. I return to this point below.

Finally, one further reason the east Asian experience is not a model for the Third World is that east Asia started its export-promotion policies at a time of relatively open markets in the "advanced" capitalist countries, and when only a few nations in the periphery were adopting such a policy. Even if this "model" could be generalized, Third World manufacturing exports would soar, which might lead to protectionism by the "First World" (see Cline 1982). This process has actually to some extent already occurred, through the renewal of the 1974 Multifibre Agreement and "voluntary export restraints" on goods produced in the periphery (Spero 1990: 85–8, 221). The World Bank has given full recognition to this problem (see World Bank 1984: 40), and criticized this, but the problem is that it lacks an analysis of *why* such protectionism takes place. It is not simply a question of an "incorrect policy" that promotes "market imperfections"; as Gamble (1983: 28) argues,

> in any actual economy the creation of rigidities and obstacles to the functioning of markets . . . [is] not accidental or the product of malign external influences but a result of the way in which markets are in fact organized. Workers will resist if their wages are driven down to subsistence. Nation-states will seek to protect themselves from unrestricted competition.

What this suggests is that "deviations" from market forces are often rooted in power relations in the world economy. These relations – which are partly constituted by market forces in the first place – are too often wished away by neoclassical theory.

Orthodox Marxism: a critique

This school of thought has been criticized elsewhere, so I will be brief here. The main problem with Warren and others is that they assume that "there is a uniform pattern of capitalist development in the Third World" (Jenkins 1984a: 39). Little attention is paid to specific class structures and struggles in the periphery and how these impact on the development of a particular country. Instead, as already shown (see chapters 2–4), these writers' sole concern is with the means by which the productive forces can be developed and so the material basis for socialism created. The problem with this approach is that contemporary developments in the periphery are assessed in terms of this long-term view, and so, once again, the selective experience of a few nations is converted into a general model.

The new international division of labour: a critique

The theory of the new international division of labour represents a systematic attempt to come to terms with some of the changes in the global economy over the past 20 years. Its main strength is that it recognizes the selective relocation to the periphery by transnational companies, and, unlike neoclassical theory, it is critical of the employment practices of many companies operating in export processing zones (see for instance Mitter 1986).

However, the question remains whether selective relocation by TNCs is significant enough to constitute a *new* international division of labour, and whether this relocation can itself explain the industrialization of parts of the periphery. The data above are sufficient to cast doubt on the principal claims of the theory, and some critics have presented at least four additional convincing reasons. First, rapid industrialization in the 1970s was eroded by de-industrialization (and a general fall in investment) in much of the debt-ridden periphery in the 1980s (Roddick 1988: 81–4, Harman 1993: 80–85). Secondly, not all of the NICs specialize in labour-intensive industries – for example, Brazil has a massive (foreign-owned) automobile industry, while South Korea has one of the largest shipbuilding industries in the world. Thirdly, there is also no necessary causal link between Third World industrialization and First World de-industrialization – job losses in manufacturing in the advanced countries are more likely to be the product of new technology or of declining demand for finished products (Gordon 1988: 39). Furthermore, the trade surpluses that OECD countries enjoy *vis-à-vis* the Third World create more jobs than are lost through imports (Jenkins 1984a: 49).

The evidence therefore suggests that this school of thought takes one tendency in the global economy and constructs a grand theory on the basis of this one tendency. This weakness is, as I have already argued, a direct product of the methodology of most theories of development and underdevelopment, whereby changes in the periphery are read off from actions in the core (see Southall 1988: 16).

This methodology is also functionalist in that the actions of Third

World states are assumed to represent the interests of transnational companies. I have already noted Frank's argument that TNCs "needed" authoritarian states in the periphery for the reconstruction of the international division of labour. He does however recognize (Frank 1981a: 274) that this is not necessarily a static situation:

> insofar as the present holders of power are able to consolidate their rule and to institutionalize their political measures, they will no longer require so much brute force to maintain themselves in power and to pursue their economic model.

The problem with this contention is that it again assumes that whatever happens in the periphery happens because it is in the interests of global capital. Political repression was required for global capital in the early stages of export-oriented industrialization and so this was "what capital got"; liberalization is a requirement of a later stage of the model, and so this too is what capital gets. In neither of these contentions does Frank even consider the notion that liberalization might be a product of social struggles from below – instead, it is simply a product of the "logic of capital". This is not to deny that capitalists, or other powerful social actors, might subvert liberalization for their own ends, but it is quite another contention to suggest that exploited classes or other oppressed groups do not enter the picture at all. As Southall (1988: 17) observes, NIDL theory "implies that labour in both the first world and the third is passive in the face of exploitation and transnational relocation of jobs".

The final methodological weakness of the NIDL school relates back to its origins in dependency and world systems theories. A central contention of this school of thought is that the NICs do not represent "proper" capitalist development, because it remains dependent on the core countries. This is the main contention of the work of Hart-Landsberg (1979, 1984), who argues that South Korea is still heavily dependent on foreign capital, foreign technology, and exports to the First World (1984: 185–8). These arguments are in some senses true, but these forms of dependence are equally applicable to many First World nations, including Britain. As Barone (1984: 195) argues, South Korea may be dependent, but this "does not alter the fact that South Korea has achieved successful industrialization". There is no harm in comparing different patterns of industrialization, but this should not be done with the implicit *a priori* assumption that country A (from the First World) is the norm and country B (from the Third World) is a shortfall from this norm, as dependency theory does (for a discussion of this point, see Mouzelis 1980). As I argued in the previous chapter, for dependency to remain a useful idea it is best utilized as an *a posteriori* rather than an *a priori* concept. As Hamilton (1986: 1, my emphasis) argues:

> International laws and interpretation of trends are developed at a high level of abstraction, but they need to be brought down to a lower level in order to be applied to the study of *particular countries* and thus the Third World as a whole.

Regulation theory: a critique

It should be clear from the above discussion that, despite protestations to the contrary (see Lipietz 1987: 4–5), regulation theory reproduces the methodological errors of dependency theory. Therefore, the criticisms applied to the theory of the new international division of labour are equally valid when applied to regulation theory. Moreover, there are further methodological criticisms that can be made.

The first point refers to Lipietz's assertion that the NIDL does not apply simply to the relocation of labour-intensive industries to the periphery, but also incorporates the rise of some capital-intensive, Fordist industry. This basically sound statement is all well and good but it does not tell us *why* capital-intensive industry has been set up; as Amsden (1990: 9) argues, "to recognize and label the emergence of more complex production in the Third World as peripheral Fordism is not to explain it". This problem is intensified by the problematic 'application of Fordism' to an inappropriate setting. The Fordist "mode of regulation" in the core was a response to the basic problem of underconsumption, which was resolved by higher wages for the workforce. However, this is not the basic problem of capitalist development in the Third World; instead, the problem is one of "*raising productivity and creating international competitiveness*, not effective demand" (ibid.: 10). This problem has been tackled in the successful NICs by a highly interventionist state, which has deliberately fostered a policy of raising foreign exchange, savings and public revenue in order to increase both productivity and competitive advantage in the international economy. These factors are neglected by regulationist work on the periphery, just as they are by neoclassical and NIDL theory, and this can again be explained as a product of a highly questionable methodological approach, which is one of "trying to understand the periphery in terms of the centre" (Amsden 1990: 9).

This problem may be derived from the weaknesses of regulation theory more generally. The empirical question of flexible production is examined in the next section, but here I want to examine one of the recurring methodological problems of this theory, namely its tendency to lean towards a functionalist mode of analysis (Bonefeld 1987: 109, Clarke 1988: 68). A stable, coherent regime of accumulation is seen as the normal state of affairs, while crises are exceptional breaks from the norm, which are functional to it in that they secure the conditions for a new mode of regulation (see Amin & Robbins 1990: 15). Social relations are relegated to a secondary place, derived from the determining structure, which is the prevailing regime of accumulation. In this way, the crisis of mass production is seen as inevitably giving way to a new structure of flexible accumulation, which will again act to serve the needs of capital. As one proponent of this thesis, Allen Scott (1988: 172), argues, regulatory forms "come into being alongside the regime of accumulation as a means of stabilising its operation through time". This leaves us with a theory that is strangely familiar; like modes of production theory, social phenomena are "read off" from the demands of the dominant, structural logic of capital. The similarities are striking: "modes of regulation" serve the functional requirements of the regime of accumulation, while "non-capitalist modes of production" serve the functional requirements of the capitalist mode of production.

Finally, and perhaps more contentiously, it could be argued that one possible implication of this new structuralism is that, in the periphery, all that is needed is the emergence of a cohesive regime of accumulation and supporting mode of regulation, and "development" will follow. In this account, the regime of accumulation can be seen as a modernizing agent, and development is reduced to the creation of the appropriate Rostowian "structural preconditions". The functionalism of structuralist Marxism therefore becomes the functionalism of modernization theory (on this point, see Leys 1986: 318).

Therefore, contrary to the arguments of Corbridge (1989: 239–

41), and despite the real insights that can be gained from an extensive examination of this school of thought, I believe that there are strong grounds for suggesting that regulation theory does not represent a particularly fruitful way for development theory to transcend its impasse. It repeats the methodological weaknesses of other development theories, including those examined in this chapter, and in particular it subordinates agency to structure, and periphery to core. Certainly it is not necessary to study *all* nations in the world in order to construct a theoretical model, as Lipietz (1987: 4–5) rightly contends, but it is necessary to study *some*.

This is a task I undertake, with special attention to parts of east Asia and Latin America, in the fourth section. Before doing so, however, I now move to an alternative analysis of the international order, which suggests that there are "laws of motion" in the world economy, which differ sharply from, say, neoclassical theory, but these "laws" should be seen as tendencies, which are subject to change by human action. In discussing these laws, I assess their implications for the industrialization of the periphery.

Industrialization, uneven development and the global economy

So far, I have identified that the common problem with each of the theories discussed is a tendency to read off the development trajectory of different countries according to "laws" identified at a global level. Put another way, the "local" is explained solely by the "global" – TNC relocation, comparative advantage or the needs of new modes of regulation. However, although I reject such a methodological approach, I do not believe that the "global" is irrelevant and that there are simply many different "localities" with their own unique development experiences. The rise of such a relativist, "post-modern" approach in the social sciences is ironic when one considers the strong globalizing tendencies (TNCs, international finance, international institutions, international communications, and so on) of the post-war years (on this point, see Archer 1991).

Instead of these two extremes, what is needed is a theory that recognizes that there are "laws" that operate in the global political economy, but that these do not operate in any mechanical way, and in particular their effects cannot be taken as explanations for the development of particular countries in the "world system". In other words, instead of the "laws" or structural needs of, for example, world systems theory, the laws of motion of the global economy should be seen as tendencies, which can be altered by human action, both within and between countries. Such a conception is therefore fully compatible with my discussion of "laws of motion" as tendencies in Chapter 2. These tendencies therefore influence, but do not completely determine, the industrial (and wider development) strategy of peripheral capitalist countries, and it is these tendencies that require explanation.

However, before moving on to such an analysis, it is first necessary to suggest some of the reasons *industrial* production is itself so important. For neoclassical theory, the need for industrialization is itself a secondary consideration; what is most important is that nation-states produce those goods in which they have a comparative advantage. As was seen above, it may be the case that some Third World societies can produce industrial goods more cheaply, and this is the explanation offered by neo-classicists for the industrialization of east Asia. However, other regions will produce other, non-industrial goods more cheaply, and so they should produce these goods and exchange them for consumer goods that are produced by industry elsewhere in the world economy. I discuss this development strategy in more detail in Chapter 6, but for the moment it is clear that, for neoliberals, industrial production is not regarded as an absolute necessity for all nations.

This view has of course been questioned by development economics since 1945. Without going into historical detail, many writers have argued that industrial production is important because it generally takes the lead in productivity increases in economies, and so has important spin-off effects in other sectors (see Kitching 1989: ch.1). Other arguments have also been put forward in favour of industrialization, such as the need for military security and the need to combat the tendency for agricultural prices to fall against industrial prices (Chandra 1992: ch.1).

One need not accept the full logic of a "super-industrializer" such as Kitching (for a critique, see T. Allen 1992) to accept that industrial production is important in assessing the prospects of any particular "national economy". Therefore, it is important that we understand how the "laws of motion" of the world economy impinge on any attempt to develop a strong and competitive industrial base in a particular Third World country. Again I stress, however, that these "laws" alone do not explain the success or failure of a particular strategy, but they are one important part of any adequate explanation.

To understand these tendencies then, I repeat the point that most capital is concentrated in the First World, and such capital flows, contrary to the implications of NIDL theory, have undergone a proportionate increase in the past 20 years. Therefore, the first task is to explain why this is the case and, conversely, why only a minority of transnational capital flows to the poor South, where labour costs, rent and taxation are usually lower than in the rich North.

This is an important question because it undermines the assumptions, and common position, of both neoliberals and the radicals of NIDL theory. Neoliberals assume that capital will flow to areas of capital scarcity because costs are lower than in capital-rich areas. This, of course, is the basis of the theory of comparative advantage and the belief that unhindered capital flows will tend to be equalized in the world economy. Although dressed in a radical rhetoric of exploitation, NIDL theory shares the same assumption – capital will flow to areas of capital scarcity because labour is cheaper (and therefore more exploited) in these areas.

To return to the question then, how does one explain the pattern of capital flows in the world economy? To understand the character and direction of these flows, one must dispense with the assumption that the world economy has a tendency to equilibrium. This assumption is derived from the equally erroneous contention that each country has the capacity to be equally competitive in the world economy. However, the competition that capitalism gives rise to does not lead to the equilibrium of neoclassical theory, but is instead based on the continuous search for surplus profits. In its quest for these profits, capital tends to become increasingly centralized, and it concentrates in particular regions, where it draws on certain "competitive advantages". These advantages include the utilization of mass production techniques, or economies of scale, such as advanced technology and skills; the organization of research & development facilities; access to credit and cheap inputs; access to the most developed infrastructures; and easy access to the most lucrative markets in the "First World" (Brett 1983, 1985, Edwards 1985). The dynamics of capital-

ist competition therefore lead to a process of centralization and concentration of capital in particular regions, and therefore to an intensification of uneven development. As Toye (1985: 10) argues:

> since the rewards of the employment of new technology will usu-
> ally contain an element of monopoly rent, it is not surprising that
> scarce factors of production like capital and skilled labour will,
> contrary to the expectations of orthodox economists, tend to be
> drawn towards areas where they are already relatively abundant.
> The equilibrating flows of capital and skills towards an area of
> relative scarcity are dominated by larger counter-flows, as succes-
> sive waves of innovation generate a continuous dynamic disequi-
> librium.

In concrete terms, what this means is that potentially competitive
capital in the periphery that tries to increase its exports to the "advanced" capitalist world faces capitals that monopolize the most
advanced technology, skills and markets and therefore enjoy considerable competitive advantages over their rivals. In an open economy,
peripheral capital faces the same problem of unequal competition
within its own economy. It is for this reason that many writers put
forward the case for state protection of domestic capital, a subject I
examine in more detail in the next chapter.

In monetary terms, this perspective contrasts sharply with the neoclassical argument that countries running balance of trade surpluses
will face increased costs as a result of an increase in their money supply. A more likely result is an expansion of credits which will facilitate
a further round of investment, including investment in more productive and efficient technology. Meanwhile, contrary to the claims of
the theory of comparative advantage, countries running a balance of
trade deficit may not automatically face a drop in prices as a result of
a contraction in the money supply. Instead, they may face a decrease
in the supply of credit, which in turn will decrease production and
lead to a rise in interest rates (see Shaikh 1980: 38–9).

So, to return to my original contention, neoclassical and NIDL
theory start from the assumption of static structures of production
(and similar conceptions of money), or at least from the view that free
trade will bring about an equalization of "factor endowments" (Ohlin
1933). What they fail to take into account is that the accumulation of

capital leads to competitive advantage on the basis of highly unequal labour productivity, both between and within sectors (Weeks 1982: 69). Therefore, just as neoclassical theory assumes that "adjustment" in deficit countries will occur through cost reduction (principally of labour), so NIDL theory assumes that relocation occurs in order to exploit cheap labour (on this point, see Amsden 1989, cited above). It is therefore unsurprising that the theory of the NIDL, and the school of thought from which it is derived, have been described as "neo-Smithian Marxism" (Brenner 1977, Jenkins 1984a). Jenkins (1984a: 34) has identified the central weakness of NIDL theory, and (as he recognizes) this criticism is equally valid when applied to neoliberalism:

There is little understanding of the role of innovation and technological development in the accumulation process. The analysis proceeds in terms of given labour processes and existing use values, while capitalism has been characterized by the continuous transformation of labour processes and the introduction of new use values.

This weakness is in turn a product of the "failure to analyse the tendency of capitalism to expand through increasing relative surplus value by raising the productivity of labour" (ibid.: 32–3).

One implication, already alluded to in Chapter 3, is that it is perfectly possible for workers receiving comparatively high wages to be more exploited than lower-paid workers, because their higher wage is more than offset by higher productivity (Bettelheim 1972). Therefore, given that surplus value extraction is higher in the core than in the periphery, it is not surprising that it is more profitable for capitalists to concentrate their investments in the First World.

Moreover, this point is not merely semantic, because it has important implications for understanding certain tendencies in the global economy. The role of the International Monetary Fund is examined in more detail in the next chapter, but here it should be noted that one rationale for its structural adjustment programmes is that cheapening labour costs will attract increased amounts of foreign investment. However, this assertion is based on the assumption of static production, which I have criticized above. Cheapening labour costs is no panacea for the Third World because unit labour costs may remain cheaper in the high-tech, high-productivity First World. The effect of cheaper labour costs, in this situation, might simply be the expansion of mass impoverishment, without any long-term benefits in terms of competitiveness.

Of course these laws are tendencies, which in turn are subject to countervailing tendencies. As Brett (1983: 95) points out, if this concentration and centralization of capital were a rigid law, then capital would now be concentrated in the hands of one single capitalist. One important means of counteracting this tendency to concentration is the action of social classes and states in resisting these tendencies, which is examined in detail below. More relevant to my discussion at this point is the continued existence of labour-intensive industries – that is, industries that have come up against barriers to increasing relative surplus value through technological innovation. The result is that labour costs still constitute a substantial proportion of total costs of production, and so in these cases the Third World might have a "competitive advantage" in low wages (although, given the persistence of low-wage sweat-shops in the core countries, even this should not be exaggerated – see Jenkins 1987: 132, Mitter 1986). This perspective, then, would partly explain the (highly selective) relocation of capital by TNCs to parts of the Third World since the 1960s. However, it should still be borne in mind that many of the exports originated from local capital and that labour was cheaper in other parts of the periphery. Nevertheless, cheap labour can still be considered *one part* of the explanation for the rise of labour-intensive manufactured industry in the periphery from the 1960s.

However, cheap labour is only one attraction for TNCs investing in the periphery. Others include the need for certain raw materials or foodstuffs, and the search for wider markets. The second of these factors might still be more important in explaining the rise of industrialization in some Third World societies. In Latin America in the mid-1970s (that is, at the height of the so-called new international division of labour), exports accounted for less than 10 per cent of total TNC output. Moreover, the share of manufacturing output in sales by US subsidiaries actually fell (from 6.2 per cent to 6 per cent) between 1966 and 1976 (Jenkins 1984b: 115–16).

The effect of TNC investment on the periphery is a matter of great controversy (see Barnett & Muller 1974, Jenkins 1987). Neoliberals regard TNCs as progressive agents promoting "even development", whereas writers of the dependency school see them as agents reinforcing global inequalities. The problem with both these approaches is that they tend to abstract TNCs from wider socio-economic structures, and so tend to see them as either the "saviour" of the Third World or an evil responsible for all of the latter's problems. Such dichotomies are again rooted in a common methodological approach that reads off changes in the periphery from changes in the core.

TNC investment should instead be analyzed as one aspect of the wider tendency towards the internationalization of capital. The effect of TNC investment in a particular Third World country will therefore vary, and it cannot be assumed that its benefits will be automatically benign (neoclassical theory) or malign (some versions of dependency theory). Instead, its effects are likely to be contradictory, and will depend in large part on how foreign capital interacts with local capital, in both the private and public sector, and with the local working class. Black and white assumptions about the effects of TNC investment are not very useful. For example, the neoliberal claim that TNCs provide employment and income to a particular country might be true in some cases, but it neglects the fact that TNCs might raise the capital within the country receiving the investment, and that raised employment in one factory might lead to the displacement of workers in another (Jenkins 1987: 96). On the other hand, dependency theory's obsession with transfer pricing may be misplaced because tax rates in the Third World are often lower than they are in the First World (Corbridge 1986: 172). This is not to deny that transfer pricing does occur (see Murray 1981), but it is not necessarily simply a means by which TNCs extract resources from the exploited periphery.

What *should* be stressed, however, is that *at the global level*, contrary to the claims of some writers (see Dunning 1981), TNCs are unlikely to alleviate the tendencies towards uneven development. There are good reasons for most direct foreign investment being concentrated in the advanced capitalist world, and it should be stressed that TNC investment in the periphery is similarly concentrated in a few countries (see Gordon 1988). *At the level of a particular country in the periphery*, TNCs might or might not intensify uneven development – this will depend on the particular social structure of the receiving country. In other words, it is not the TNCs *per se* that promote uneven development, but the wider social structure of which TNCs are one part. This again brings me back to my point that this social structure cannot be read off from the needs or whims of global capital, but must instead be the subject of careful empirical analysis. To illustrate this point, I undertake a brief but suggestive analysis of the east Asia experience below.

However, before moving on to this section, one final argument should be examined, as both a conclusion and defence of this section. My key argument has been that uneven development in the world economy is a product of the competitive accumulation of capital, and in particular capital's constant search for surplus profit. Historically, this has laid the basis for a tendency towards economies of scale, whereby unit costs are reduced by increasing production size, firm integration, and so on. Recently, however, this focus on mass production and its developmental implications has been challenged by approaches that emphasize the efficiency of flexible production methods. This was discussed above in the context of regulation theory, and some writers (Urry 1989) have linked increased flexibility to the alleged rise of the "new international division of labour". This argument has already been criticized above, but there is another argument, which I have not dealt with. This is the argument of the "flexible specialization" school, which argues that capitalism has entered a new period in its development, based on small-scale, efficient production, utilizing the latest microelectronic technology, and serving small, specialist niche markets (Piore & Sabel 1984). The implications of this argument for the industrialization of the periphery have not yet been fully scrutinized, but the implications of the decline of economies of scale (or at least those scale economies internal to the firm) should be clear: the decline of mass production has opened up new possibilities for the rise of competitive industry in the periphery. Initially, Piore & Sabel (1984: 279) put forward the view that flexible production in the core countries might be accompanied by the rise of mass production in the periphery. Sabel (1986: 43) has since developed this argument and put forward the idea that the periphery itself can adopt flexible specialization techniques, and so compete effectively with the core countries. Moreover, he argues that the Third World (his focus is however on Latin America) has certain advantages in adopting these techniques: first, "poor organization" (from the viewpoint of the mass production paradigm) in existing large factories in Latin America has meant that within a factory there may in effect be independent producers, specializing in small-scale craft production; and secondly, there is the "informal sector", which may form the basis of a cluster of small-scale independent producers, adapting to volatile markets and new technology (ibid.: 46–8, see also Institute of Development Studies 1987, Rasmussen et al. 1992).

The basis of this optimism is the rise of new industrial districts, uniting small-scale producers (and so sustaining external economies of scale) in the core countries. The most commonly cited example is the "Third Italy", based on small-scale production, most notably in the machinery and textiles sectors (Sabel 1982), which is said to serve specialist markets. However, the arguments that mass markets have ended and the need for internal economies of scale no longer exists is highly questionable (Williams et al. 1987: 424–9) – indeed, the work of the flexible specialization school is characterized by a paucity of evidence on this key central point. The supposed break-up of mass markets should not be taken as a sign of a new "post-Fordist" capitalism, precisely because it is far from comprehensive. Indeed, in so far as any collapse of mass markets has occurred, this should be seen as a *consequence* of the persistence of high unemployment, unequal income differentials, and so on (Gough 1986: 63).

The persistence of small-scale production should be seen in a similar light. The flexible specialization thesis is essentially "dualist" in that it expects one form of production (flexibility) to replace another (mass production). In fact, the two are closely related and it is often the case (as regulation and NIDL theory, albeit one-sidedly, point out) that "Fordist" production processes also utilize, and subordinate, small-scale production (Amin & Robbins 1990). This is often reflected in poor working conditions and long hours in small factories, which, contrary to Piore and Sabel's optimistic claims, actually seems to be the case in the "model" of the "Third Italy" (Murray 1987, Amin 1991).

Finally, and most importantly for my wider discussion in this section, the "flex.-spec." school's analysis is based on a highly technicist account of economies of scale. This repeats the errors of neoclassical theory, which regards economies of scale as the product of a particular technology, such as heavy industrial machinery. However, economies of scale actually emerged as a strategic consideration of individual firms in the context of the competitive accumulation of capital. Firms integrated through internalizing the supply of key inputs, and thus dismissing market-place uncertainty. Moreover the expense of this process precluded smaller firms from following a similar strategy. It is in this social, rather than purely technological, context that economies of scale were developed, and it is in this social context that economies of scale persist. As O'Donnell & Nolan (1989: 9) argue:

> Economies of scale thus emerged as a key dynamic in the competitive process *after*, and not *before* significant increases in capital concentration and centralization . . . Just as the historical evidence points against the view that the giant corporation was a creature of technology, so it is unlikely that it is about to wither away under the impulse of a new wave of technological developments.

So, my discussion of economies of scale retains its validity, despite the optimistic contentions of the flexible specialization school. Whereas theories such as underdevelopment and NIDL theory may be deemed excessively pessimistic for regarding power relations in the world economy as insurmountable, the flexible specialization thesis is excessively optimistic because it simply wishes these power relations away. Sabel's concern with the informal sector in Latin America is one thing; but his implied contention that it can compete with large Western transnational companies is something else. What the 1980s and 1990s have witnessed is a continuation of these unequal social relations in the world economy, and indeed in some respects an intensification of economies of scale (Amin & Robbins 1990: 25–9). On the other hand, the persistence and intensification of uneven development in the world economy do not automatically mean the persistence and intensification of a simple North–South, industrial/non-industrial, divide. I elaborate on this point below.

Industrialization in east Asia

The discussion in the previous section was not intended to be an explanation for the industrialization of selected Third World countries since the 1960s. Instead, it was a critique of NIDL and neoclassical accounts of the *global economy*, and of their implied account that it tends towards equilibrium. In presenting an alternative, I suggested that the "laws" that do exist in the world economy should be regarded as tendencies, rather than simple rules that affect countries in a particular way. Therefore, the specific effect of these tendencies on a particular country (or group of countries) in the world economy will depend on how that nation and its institutions (especially the state) and social actors respond to these "laws". The economic performance of Tanzania, for example, cannot be read off from the laws of the world economy, despite the claims of true believers in both the Third Worldist and neoliberal camps.

This section therefore concerns itself with how parts of east Asia (particularly South Korea and Taiwan) have industrialized over the past 30 years. The first point is one that I have already alluded to in my critique of neoclassical theory above, which is that the state has been very active in the promotion of industrialization in *both* regions. The figures cited above make that clear, and they may even suggest that, if anything, the state was more active in promoting industrialization in east Asia than in Latin America. Given my discussion of the tendency of capital to concentrate in particular regions and maximize profitability through economies of scale, such a high degree of state intervention is hardly surprising – this is the most effective means of protecting domestic industry from foreign competition, and (at least in the case of east Asia) it is the best way of controlling capital flight.

So, as I argued above, it is not state intervention (or lack of it) that itself accounts for the superior economic performance of east Asian NICs as against Latin American NICs; rather it is the character and quality of that particular intervention. For instance, legislation in Korea passed in the 1960s stipulated that any illegal overseas transfer of $1 million or more was punishable by a minimum sentence of 10 years' imprisonment and a maximum sentence of death (Amsden 1989: 17), which can hardly be accurately described as a "minor intervention" in the free market. Such stringent controls did not exist in Latin America, where capital flight in the 1970s and early 1980s (i.e. before the so-called debt crisis) was an endemic problem (Sachs 1984: 395). As I showed above, subsidies given to industry in east Asia were often tied to export performance, but no such policy existed in Latin America and so the state was left protecting high-cost, inefficient producers, without any compensation in terms of breaking into world markets and generating foreign exchange.

However, merely to point to the efficiency of the state in one region and not the other is not to explain the reasons for these discrepancies. To leave the analysis at this point would leave us with the kind of "state fetishist" analysis that I criticized in Chapter 3. What is therefore needed is an analysis that situates state policies in a *social context*, and thus avoids the mistake of seeing the state as a purely technical instrument. As Jenkins (1991: 200–201) shows:

> The question is not so much how to explain policy differences, but rather how to account for differences in the effectiveness of economic policy in the two regions . . . [Such a focus] implies that the failures in Latin America are not a consequence of incorrect policy choices, but of the way in which the state's lack of autonomy precluded certain policies from being pursued.

What is therefore needed (and once again this takes us "beyond the impasse") is a comparative analysis of the social history of the two regions. A full account of the specificities of the two regions, and of countries within these regions, is beyond the scope of this chapter (for more detailed, but still schematic surveys, see Evans 1987, Jenkins 1991). What I can do here is suggest at least some of the reasons for the character of state intervention and industrialization. In examining these differences, the focus is on the "success stories" of east Asia, and in particular Taiwan and South Korea, but brief "suggestive contrasts" (Runciman 1983) will be made with Latin America. Although there are many other factors, I will concentrate on three – geopolitics and the Cold War, land reform, and the character of the state *vis-à-vis* civil society. Once again, though I risk repeating myself, it should be stressed that the purpose of this comparison is not to put forward east Asia as a model for Latin America; it is simply to explain two different industrialization paths.

Taiwan and South Korea had particularly turbulent post-war histories, and both nations were products of the defeat of the Japanese colonial power in 1945 (which is itself significant – see Cumings 1987: 45), and of resistance to the rise of Asian communism. The state of South Korea was a product of the division between the communist North and capitalist South, and Taiwan was in effect taken over by the defeated nationalists from the Chinese mainland after communist victory in the latter in 1949. It is therefore unsurprising that both states played a key role as defenders of United States interests in the region during the Cold War. Both states were regarded as bulwarks against "communist expansion" in the region, and so both states were highly favoured by successive US administrations in terms of the volume of aid that they received. From 1946 to 1978, South Korea received nearly $6 billion in economic aid, while Taiwan received some $5.6 billion. This compares with $6.89 billion for the whole of Africa over the same period. Needless to say, the amount of military aid was even greater (Cumings 1987: 67).

The importance of aid is most clearly quantified by the fact that it financed around five-sixths of South Korea's trade deficit in the 1950s (Cumings 1987, Bello & Rosenfeld 1992: 4). However, the importance of aid alone cannot explain the differences in industrial performance in the two regions, because per capita income in Korea and Taiwan remained far lower than in Argentina and Mexico in 1960, when levels of aid started to fall. Nevertheless, high levels of aid were important in financing infrastructural developments in the region in the 1950s.

More important however was land reform. Once again, land reform must be rooted in the turbulent politics of the immediate postwar period. In Taiwan, fearing a mainland-style peasant rebellion, the Chinese nationalists (Kuomintang) introduced a radical land reform. This would have been unthinkable in mainland China because their supporters were mainly landowners, but this was not the case in Taiwan. The result of this land reform was that "almost overnight the countryside in Taiwan ceased to be oppressed by a small class of large landlords and became characterized by a large number of owner-operators with extremely small holdings" (Amsden 1985: 85). Although land reform was not as far reaching in South Korea, it still followed a broadly similar path, and so in both cases a powerful landlord class had been expropriated and a smallholding peasantry established. The land reforms themselves were a kind of "punitive forced investment in industry" (Selden & Ka 1988: 115). The landlord class was forcibly turned into peasants or workers through bankruptcy, or became a capitalist class, under the close leadership of the state, who received shares in government monopolies (ibid.). Moreover, the newly established peasantry now faced a direct relationship with the state rather than a class of private landowners (Wade 1983), and in the 1950s and 1960s "primitive accumulation" was secured as the terms of trade were fixed (by the state) against agriculture (Selden & Ka 1988: 123–6), and by a steady flow of labour, which migrated to the cities and thus provided industry with a labour force (Hamilton n.d.: 156).

The implications of this land reform for future industrial development were enormous. It enabled the Korean and Taiwanese states to extract a surplus from farming to finance industrial development, without leading to an absolute impoverishment of the peasantry. The state provided new technology, credit and an educational and scientific infrastructure and ensured that these benefited the majority of the peasants. Once the technology was dispensed, productivity increased and the surplus was sold at low prices to the state. This process ensured a surplus from agriculture, which helped to finance industrial development without leading to an absolute impoverishment of agriculture.

There are strong grounds for suggesting that this process of land reform is unlikely to be repeated elsewhere, principally because of the specific circumstances in which land reform was introduced in Korea and Taiwan (Byres 1991: 64). Thus, in the case of east Asia, agriculture supplied industrial capital with a labour force, a surplus, foreign exchange (through agricultural exports), demand for industrial output (such as fertilizers and machinery) and a market for consumer goods. This is in marked contrast to the situation in much of Latin America (as well as Africa and the rest of Asia – see Williams 1988, Hamilton 1987, Byres 1991), where, despite land reform, large land-owners continue to exercise strong influence over the government and to extract absolute rent over an impoverished peasantry, many of whom have to try to find additional work in the already "overburdened" cities (de Janvry 1987).

Finally, and closely related to this last point, is the relationship of the state to civil society. The development of class forces and interests in east Asia has been closely tied to the development of the state, whereas classes in Latin America have been far more powerful *vis-à-vis* the state. This has given the state in Korea and Taiwan a far greater degree of "autonomy" from social classes than is the case in Latin America. The continued power of an unproductive landowning class in Latin America has already been noted, and this has hindered the "room for manoeuvre" of states in that region. The power of the state in east Asia also explains how it made the switch from import substitution to export promotion fairly easily compared with Latin America, where the bourgeoisie and the working class successfully resisted the removal of protectionist measures and maintained a longer allegiance to ISI (Jenkins 1991: 208). It is also for this reason that the east Asian states were far more selective in their openness to foreign capital investment than was Latin America (Wade 1990: 363–4).

It is clear, then, that there are specific social reasons for the industrial development trajectories that have been followed in the two regions, and these are rooted in the development of a specific class structure and a specific state form. It is in this respect that it can be argued that *culture* is central to the process of capitalist development. This does not entail a cultural reductionism whereby east Asian culture, abstracted as an homogeneous Confucianism, explains development (see Bauer 1984a: 86–7, Berger 1986). Instead, culture refers more widely to the specific social history of the region, and the social relations that emerged out of this history. These were the key factors that shaped the character of Korean and Taiwanese industrialization. The Western stereotype of Confucianism promoting industrialization leads one to ask why industrialization did not occur earlier; and the argument that Confucianism promotes trust, loyalty and stability simply flies in the face of the highly unstable social histories of Taiwan and South Korea since 1945 (as well as the rest of the region) (see Henderson & Appelbaum 1992: 15–17, Berry & Kiely 1993).

One final brief point about the state in east Asia: "state autonomy" should not be conflated with a "political determinism" whereby "the political determines the economic". I criticized the conceptual separation of economic and political levels in the previous chapter, and showed how the appearance of separate instances was rooted in the development of capitalist social relations. The Korean and Taiwanese cases show quite clearly that the development of capitalist industrialization is rooted in the development of a specific state that has its roots in the social struggles and transformations of the post-war years. The state has played a leading role in the accumulation of capital, and its "peculiar" role (and one should again bear in mind my point that all development trajectories are in some sense peculiar) is rooted in class struggles in the region. Therefore, it is a case not of the state having an "autonomy" so much as being a constituent part of, and contributor to, social struggle and social transformation (for an interesting account of state "autonomy" that also rejects the separation of the economic and the political, see Gulalp 1987).

Conclusion

Mainstream accounts of Third World industrialization suffer from a tendency to explain the phenomenon in solely global terms, which leads to the creation of new "fetishized" accounts or models of social reality. On the one side, the radicals argue that the rise of selected NICs is part of a constant process of Third World subordination to the Western world – in this case, to the needs of Western capital. Thus, a continued and never-changing model of the world system and under-development is posited. The implication is that there is no need to look at events *within* the periphery, because these are *de facto* not "proper" developments and are simply serving the needs of Western imperialism (see Frank 1981a,b, Hart-Landsberg 1984). On the other side, the free marketeers argue that the rise of selected NICs is a product of the adoption of free market principles operating at an international level. Thus, a model of adopting free market principles is posited for the rest of the Third World. Although this does not preclude analysis of events within the periphery, this school of thought does so only to the extent that particular nations embrace the global force of comparative advantage, and that governments adopt the correct policies. The continued search for these two principles, however, precludes an adequate analysis of how the state might modify or completely challenge the principles of comparative advantage, and it reduces the policy-making process to a purely technical discipline. Therefore, "open markets" become the medicine for the rest of the world, which conveniently leaves aside the fact that Korea and Taiwan never adopted this medicine in the first place, and that other countries have hugely different social structures and relations (see Hamilton 1987).

The alternative, and vastly less ambitious, approach adopted here is that there are strong tendencies operating in the global economy that militate against the rise of the Third World (and, I argue in the next chapter, of the former Second World) countries as competitive industrial nations. This does not preclude *some* countries, including "late developers", from becoming rapid industrializers, but they are hardly likely to do so on the basis of adopting free market principles. Whether or not they become *successful* (and this term itself needs to be deconstructed – for useful popular surveys of east Asia, see Ogle 1990, Bello & Rosenfeld 1992) will depend on specific social struggles and relations within that particular country, not on "correct policy" or a passive subordination to Western interests.

6

The politics of the impasse I: states and markets in the development process

The neoliberal "counterrevolution" is perhaps the most important factor in development studies over the past 15 years. At the same time, however, it has also been neglected, particularly by those scholars whose work is derived from a sociological tradition. The previous chapter provided the basis for a critique of the neoliberal paradigm, but confined this to an assessment of east Asian industrialization. This chapter extends that critique by looking more generally at the role of states and markets in the development process, and briefly applying my discussion to the development potential of post-communist societies.

First, I examine the nature of the neoliberal paradigm, and its belief that developing societies can (and must) rapidly develop on the basis of adopting policies of limited government and therefore allowing market forces a free hand. Secondly, I question this approach by returning to my discussion in the first part of the book, and suggest that neoliberalism *fetishizes* both the state and the market. Thirdly, I draw on this discussion to suggest that the neoliberal model has unexpected consequences when actually applied to concrete social structures in the periphery. On this basis, I provide a critique of neoliberalism in both theory and practice. In so doing, some of the key "buzzwords" of development institutions – good governance and democracy – are examined. Fourthly, I extend this discussion by questioning the claims that neoliberal modernization theory has made about post-communist societies.

The neoliberal paradigm

As was shown in the previous chapter, the return to the principles of comparative advantage, unregulated free trade and limited government was rooted in the failure of state-led development strategies, and in particular import-substitution industrialization. The neoliberal claim that the Third World could rapidly develop by rolling back the state and competing in an open world economy was (supposedly) further reinforced by two factors – the rise of the east Asian NICs and the collapse of state socialism in the late 1980s and early 1990s. The first of these factors was addressed (and challenged) in the previous chapter; the second is examined in the following chapter.

What is indisputable however is that state intervention in many parts of the periphery *was* inefficient and counter productive. Neoliberals argue that this is the case for three principal reasons: first, state intervention leads to the protection of inefficient economic activity; secondly, it leads to "rent-seeking" activity by specific interests both within and beyond the state; and, closely related to this point, thirdly, it leads to discrimination against certain economic agents, such as food producers. Each of these points requires elaboration.

The first point is that state protection to domestic producers leads to consumers in effect subsidizing inefficient producers, and therefore having to pay for expensive goods in the market place. This process is reinforced by other kinds of state intervention in the market place. For example, a state-protected and overvalued exchange rate leads to discrimination against exporters and encourages cheap imports (World Bank 1984: 35). As the previous chapter showed, this was one of the main criticisms made by neoliberals of import-substitution industrialization.

Secondly, state intervention leads to "rent-seeking" activity. Such rent-seeking can be broadly defined as unproductive income-earning economic activity derived from state regulations (Kreuger 1974, Bauer 1984a: 83–4). For example, controlling imports through licences leads to competition for these same licences. Such competition in turn leads to lobbying or corruption, rather than productive economic activity. Similarly, high state tariffs in one sector will lead to higher prices in that particular sector, so too many entrepreneurs will move into this more profitable venture. The blame for such market distortions is placed firmly at the door of the state, or "the politicization of (economic) life" (Bauer 1984b: 33–4). The World Bank (1988: 35) has nicely summarized this view:

> Although the pursuit of private interests allocates resources efficiently in competitive markets, this generally does not occur when governments use the monopolistic powers of government to their own advantage. Politicians, bureaucrats and many private interests gain from a growing government and greater government expenditure.

Thirdly, and following on from this point, the state discriminates against some sectors at the expense of others. Much is made of the low prices paid to domestic food producers in sub-Saharan Africa and the impact that this has had on domestic food supply (World Bank 1981: 55). According to the World Bank (1983: 53), "[p]rices set low to benefit consumers – especially for food – have frequently discouraged producers, creating scarcities and greater dependence on imports". State marketing boards in Africa often set low prices for farmers' products, with the result that the state-controlled market was often by-passed and illegal markets developed in their place. These in turn encouraged corruption as state officials "turn a blind eye" to the operations of this market (another form of rent-seeking).

So, to summarize: the neoliberal paradigm, which strongly influences international institutions such as the IMF and the World Bank, asserts that the economic problems of the developing world can be attributed to too much government and a failure to allow market forces to operate properly. The proposed remedy is therefore the encouragement of the private sector and the liberalization of "national economies". In order to fulfil these objectives, three key policy proposals are recommended: currency devaluation, limited government and incentives to the private sector, and the liberalization of international trade.

It is argued that devaluation of the national currency encourages exports by making them cheaper and discourages excessive imports by making them more expensive. In this way, growth through exports is promoted (the "Korean model") and constant balance of payments deficits are resolved (World Bank 1984: 35).

Secondly, the promotion of limited government entails the removal of too much government interference in the economy, which crowds out private investment and leads to market distortions. In practice, this would include the end of price controls, cutbacks on state subsidies such as cheap food, the ending of minimum wage legislation, and full-scale privatization. More recently, the World Bank (1989, 1992) has argued that there also needs to be some kind of focus on the institutional framework that will enable the reforms to be carried out (see below). Neoliberals accept that markets do not always work perfectly, but they argue that market distortions are preferable to state imperfections (Lal 1983: 106). Thus, even when markets are only a second-best welfare option, they are preferable to state intervention because the cost of this intervention outweighs the benefits. State intervention is therefore deemed to be more inefficient than development led by market forces.

The third proposal is the liberalization of international trade. In practice, this means the end of import controls and the removal of excessively high tariffs. It is argued that these policies will force domestic economic actors to be efficient because they will now face competition from foreign companies. This contrasts with import-substitution development strategies, which led to the protection of inefficient producers from foreign competition.

Taken together, these policies enable economies to pursue their comparative advantage – that is, they enable countries to specialize in those goods that they produce cheaply and efficiently. This practice can be discovered only via a process of free market competition because protection ensures that inefficient economic activity will persist. If all countries operate according to these free market principles, and therefore exercise their comparative advantage, then all countries can benefit through participation in the world economy, so that country A can exchange the products of its comparative advantage with the products of country B's comparative advantage.

This then is the basic case for market forces as made by proponents of the development counterrevolution. Before moving to an assessment of the adequacy of the neoliberal case, I first make a brief detour and examine the separation of state and market, which lies at the heart of the neoliberal case. This examination is relevant because, like so much development theory, neoliberalism naturalizes historical and social phenomena – in this case, the separation of a political state from an apolitical, impersonal economic sphere known as the market place. In showing that this separation is not natural, but is the product of specific historical factors, I advance the argument in the third section by explaining how the imposition of a natural model on concrete social relations has consequences that are not accounted for by the free market model.

The "separation" of state and market

At the heart of the neoliberal perspective, then, there lies a duality between the state and the market. The role of the state as a development agency is simply to provide the right framework for market forces to flourish. This enabling role would include the preservation of law and order, the guarantee of private property and contract, and the provision of some "public goods" (although there is much dispute over what precisely is a public good). Thus, according to Adam Smith (1910: 180–181):

> the sovereign has only three duties to attend to . . . first, the duty of protecting society from the violence and invasion of other independent societies; secondly, the duty of protecting, as far as possible, every member of the society from the injustice or oppression of every other member of it, or the duty of establishing an exact administration of justice; and thirdly, the duty of erecting and maintaining certain public works and certain public institutions, which it can never be in the interest of any individual, or small number of individuals, to erect and maintain.

However, I will contend in this section that such a separation is based on a false dichotomy between state and market, whereby the two separate spheres are assumed to be natural. Instead, it will be argued that the "separation" of the two is historically and socially constituted, and so cannot be assumed as natural.

The appearance of separate political and economic spheres is unique to capitalist social relations, and in particular those relations that emerged in England. The full implications of this argument will be drawn out in the next section, but it can immediately be noted that, if the state and the market cannot be separated as such, then it must follow that the latter needs the former, and so analysis should focus on the relations between the two.

Most Marxist analyses of the state have failed to conceptualize the "internal relations" (Ollman 1976) between state and market, and instead have, like the neoliberals, fetishized the separation of the two. Analysis has often taken the form of an economic base determining a political superstructure, so that the state is relegated to a derivative political sphere and reduced to a technical machine made up of armed bodies of men (*sic*.) (Lenin 1977: 267–70). More subtle analyses have attempted to theorize the capitalist state in terms of how it serves, or remains a prisoner of, the needs of capital (Poulantzas 1973).

The problem with these approaches is that they take for granted the immediate appearances of capitalist social relations, and fail to see the inner connections between the "economic" market and the "political" state (Clarke 1977, Holloway & Picciotto 1978, Meiksins Wood 1981, Gulalp 1987, Burnham 1993). In this respect, these views simply repeat the errors of neoliberal theory. Marx (1977: 30, my emphasis), on the other hand, argued that "the abstraction of the *state as such* belongs only to the modern time . . . The abstraction of the political state is a modern product." This abstraction is closely tied to the development of a civil society of independent producers, competing in the market place. Again Marx (1977: 56, my emphasis) is quite clear on this point:

> The *formation of the political state* and the dissolution of civil society into independent *individuals* . . . is completed in *one and the same act*.

This statement should not be taken in the literal sense that the "separation" of state and economy automatically produces capitalism (Sayer 1991: 72–3). Instead, the "separation" of economy and polity should be regarded as a form of appearance of capitalist social relations (and, it may be argued, a form of appearance far more marked in the case of English, as opposed to French, capitalism; see further Meiksins Wood 1991, Sayer 1992). This point is best illustrated by a comparison of capitalist and feudal societies.

Within feudal societies, there is no separate political sphere as such. Political and economic power are unified in the hands of particular estates, and social identities are inseparable from membership of these estates. This scenario applies equally to trade guilds as much as to serfs – in this sense, feudalism can be identified as "the democracy of unfreedom" (Marx 1977: 30). Membership of a particular estate was identical with membership of "the state".

Feudalism entered a period of crisis in England from the fourteenth century, which in turn resulted in the establishment of the absolutist state (Anderson 1974: 113–42). This state was an attempt by the ruling class to defend its privileges, but in the long run it had the effect of concentrating power, and thus facilitating the rise of capitalism. The "parcellized sovereignties" (Anderson 1974: 19) of feudal society were gradually broken by the emergence of an impersonal sovereign body, which thereby undermined the relations of personal dependence that characterized feudalism. Therefore, it was precisely the emergence of the political state that laid the basis for the emergence of the abstract individual, freed from the bonds of personal dependence on the estate. It is in this regard that Durkheim (1957: 64) argued against Spencer that "it is only through the state that individualism is possible".

The long development of the English state therefore secured the conditions of the development of capitalist social relations. For instance, in the sixteenth century the Tudor revolution in government established national sovereignty, and there was widespread enclosure of land. The latter half of the seventeenth century saw the breaking of the independent power of the Crown, the establishment of the Treasury and the Board of Trade, and the reconstitution of common law as a guarantor of contract and property ownership (Corrigan & Sayer 1985, see also Meiksins Wood 1991: 45–54).

The implications of this brief discussion for an assessment of neoliberalism are enormous. The origins of the "wealth of nations" lie not in a natural tendency to truck, barter and exchange, so much as in the development of capitalist social relations in which the state and the market are intrinsically connected. Indeed, it was precisely the development of the political state in England that helped to facilitate the generalization of a "barter and exchange" economy. In this respect, it is far more accurate to suggest that commodity production was generalized in England *because of the state* than it is to argue that it was generalized in spite of the state.

The inadequacy of the neoliberal conception of state and market should now be clear. It takes the form of appearance of capitalist social relations in England and regards them as a model for the twentieth-century developing world. This is also clearly inadequate as a model of the rest of the "advanced" capitalist world, where even the appearance of a separate economic and political sphere hardly existed (Gerschenkron 1962). However, it is the model of free market capitalism for the developing world that is relevant here.

States and markets in the development process

I now draw on the argument of the previous section to criticize the attempted implementation of an ahistorical model by the World Bank and the IMF in the developing world. This is done by dividing this section into three parts: first, an examination of World Bank and IMF policies and their unexpected (and largely negative) consequences; secondly, an assessment of the World Bank's own explanation for what has gone wrong, and an alternative interpretation; and thirdly, following on from the argument in part two, an examination of the prospects for liberal democracy in the developing world in the 1990s.

Structural adjustment programmes and unexpected consequences

The above discussion makes clear that the apparent separation of state and market is rooted in the immediate appearance of capitalist social relations. Neoliberal analysis confines itself to these appearances and argues that the problems of the economies of the developing world are the product of "too much government" (World Bank 1981: 4.1). However, given the intimate connection between state and market, it follows that the attempt to weaken the state and at the same time strengthen the market might have unexpected effects. This observation is reinforced by the closely related fact that the neoliberal focus on markets abstracts from the power relationships on which market forces are based. People participate in real markets on highly unequal terms, based on unequal ownership of resources, which in turn may facilitate power over the ways in which markets operate (Mackintosh 1990: 50–51). For instance, United States policy has had a major impact on the structuring of international food markets through the dispensation of food aid (Friedmann 1990). Mackintosh (1990: 50) has ably summarized the basis of this alternative view:

Better models recognize that markets concentrate information, and hence power, in the hands of a few: that some participants are "market makers" while others enter in a position of weakness; that markets absorb huge quantities of resources in their functioning; that profits of a few, and growth for some, thrive in conditions of uncertainty, inequality and vulnerability of those who sell their labour power *and* of most consumers; and that atomized decision-making within a market can produce long term destructive consequences – for example on the environment – which may have been intended by none of the participants. Finally, there is no such thing as a free market: *all* markets are structured by state action; the only variation is how the terms of their operation are set.

It follows from this approach, then, that structural adjustment programmes may actually have the effect of increasing the vulnerability of the weakest, who can no longer afford to buy previously subsidized food or receive educational or health provision. Even "getting prices right" in agriculture may backfire, as food producers, who are supposed to benefit from such a policy, may find that increased prices for their products are more than counteracted by an increase in prices for the essential goods that they have to buy (Bernstein 1990a, Harriss & Crow 1992).

Moreover, unequal resource endowments (Sen 1981) are gendered as well as "classed" and so it is not surprising that structural adjustment has often hit working-class women the hardest. An increase in labour force participation (which may be in the informal sector) by women is common as working-class families attempt to make up for wage cuts, but this means that in highly gendered societies women are expected to endure the burden of work in the cash as well as the domestic economy (Young 1993: 36–9). Alternatively, the burden of women's production for direct use may increase in the context of falling real incomes (Elson 1992).

It is also the case, as the previous chapter made clear, that capitalism reinforces hierarchies *between* as well as *within* countries. This is an important point because, rather than fulfilling the World Bank expectation that the liberalization of international trade will enable countries to fulfil their comparative advantages, the result may be the intensification of these hierarchies (Shaikh 1980). Already established capitals have the benefit of the most advanced technology (and therefore higher labour productivity), a highly developed infrastructure, and access to well-established, lucrative markets. "Late" developers therefore face serious *competitive disadvantage* in breaking into the established markets of the advanced capitalist world, and for similar reasons local capital in the developing world faces unequal competition from cheap imports originating from those same "advanced" capitalist countries. These problems are not necessarily insurmountable, as the previous chapter made clear, but they do suggest that liberalization is hardly a panacea for the developing world. Moreover, as the previous chapter demonstrated, it is likely that, no matter how great the incentives supplied for foreign capital to invest in the developing world, these alone are not sufficient to overcome the benefits to capital of continuing to concentrate investment in the metropolitan capitalist countries (see also Jenkins 1984a, Kiely 1994). This observation is confirmed by recent statistics. Of the $150 billion of foreign investment in 1991, more than two-thirds went to the industrial capitalist countries (excluding the NICs). Moreover, nearly 70 per cent of the foreign investment that goes to the developing world ends up in just ten countries, which include the NICs (United Nations Centre on Transnational Corporations 1992, *Fortune* 1992, *New Internationalist* 1992). So, if it is accepted that there are good reasons for arguing that foreign investment is likely to remain concentrated in selected countries in the world economy, then the effect of structural adjustment programmes is likely to be the intensification of poverty without any corresponding benefit in terms of a significant increase in foreign investment.

This discussion then shows some of the problems with a "market-led" approach to development in the Third World. Even the IMF "success stories" of the 1980s – Botswana, Ghana and Mexico – have actually had very mixed economic and social results, and the view that these "successes" have been due to "correct policies" is at least questionable (Laurell 1992, Kaplinsky 1993, Sandbrook 1993). The International Monetary Fund (1989: 44) has itself been forced to admit that "the implementation of policy reforms often was less successful than expected, and the increase in financing from private sources did not materialize". However, the neoliberals at both the World Bank and IMF have a particular interpretation of the failure of structural adjustment.

The failure of structural adjustment: good governance versus social relations approaches

The question that must now be addressed is *why* have structural adjustment programmes largely failed? The World Bank itself has its own explanation for failure, which argues that the medicine itself was not wrong, but that the institutional framework within which structural adjustment was implemented was inappropriate. From the late 1980s, the Bank has argued that what is needed, particularly in sub-Saharan Africa, is "good governance" (World Bank 1989, 1992). Although there is some disagreement over what constitutes good governance (the World Bank for instance is less concerned with democracy than are some proponents), Linda Chalker, the British Minister of Overseas Development in 1992, provided a useful summary definition:

> This means accountability and transparency in the decision-making process. It means political pluralism with free and fair elections. It means the rule of law and freedom of expression. It means far less spending on military hardware and war-making and much more on primary schools and healthcare. It means fighting the cancers of graft and nepotism. (Cited in Moore 1993a: 3)

The question of democracy is examined in detail below, but what is clear is the emphasis on public accountability, pluralism and the rule of law, which are also stressed by the World Bank (1992). What is less clear is how these practices should be achieved without a heightened degree of participation by lower classes in the developing world. The World Bank (1989: foreword) has even recognized the truth of this statement by paying lip-service to the concept of empowerment. However, the Bank is still largely committed to the notion that the adoption of market forces is the best route to economic growth and, moreover, to the empowerment of "ordinary people". However, as the discussion above suggests, attention to markets is just as likely to increase some people's vulnerability as it is to empower others. This point relates to the neoliberal conception of democracy, which is examined below.

First, some comment should be made about the notion of good governance as a means towards economic growth. To the World Bank's credit, the emphasis on governance represents something of a move beyond a narrowly economistic approach in which markets are abstracted from states (see the previous section), but there are grounds for arguing that the World Bank has not gone far enough. Governance is still largely measured according to its ability to promote market forces, and so the notion of separate economic and political spheres remains intact. This is clear from the statement that what is required is "not just less government but better government – government that concentrates its efforts less on direct interventions and *more on enabling others to be productive*" (World Bank 1989: 5, my emphasis). The state is therefore reduced (by the *analysis*, but not necessarily by the *reality* – see the discussion of democracy below) to playing a purely enabling role for the economy, where the really important events are deemed to occur. This rests on an ahistorical, fetishized account of the construction of capitalist social relations in England, where the "separation" of state and economy rested on contingent *social factors*.

This Anglo-centric approach to the state is very different from the reality of late developers, not least the most successful recent late industrializers in east Asia, as the previous chapter showed (see also Moore 1993b). Given the hierarchies that exist in the world economy discussed above, it seems likely that any future "late" capitalist developing nations will similarly rely on a highly interventionist state. This is not to deny that many states in the developing world are inefficient, but this must be demonstrated rather than theorized on an *a priori* basis devoid of any historical or social content.

Such an emphasis on *social relations* provides the basis for an alternative account of the results of structural adjustment programmes. These programmes are not simply the products of technical economic policy, but are implemented in concrete social environments with specific social relations. The reality of "separate" economic and political spheres (that is, a limited government existing side by side with "unhindered market forces") should not be naturalized into an ahistorical model, but should instead be grounded in specific social histories. The imposition of free market policies on very different social structures has had very different effects from the expected operation of comparative advantage and rapid export-led growth. For instance, marketization may simply reinforce supposedly non-capitalist productive activity such as unfree forms of labour, or increase the unproductive activity of merchant capital, which profits from trade on the basis of the preservation of "non-capitalist" productive activity (Watts 1990, Raikes 1988: 47–8, Burawoy & Krotov 1993, and the fourth section below). In either case, contrary to the hopes of the neoliberal model, there is little compulsion for the owners of the means of production rapidly to develop the productive forces and substantially boost economic growth. This is not to say that neoliberal policies will automatically promote stagnation in the Third World (the under-development position), but neither will structural adjustment automatically promote a rapidly expanding capitalist economy (a new version of the inevitabilist position – see Chapter 3).

Therefore, the promotion of an efficient capitalism in the developing world should be seen as less the outcome of technical policies (such as those of the IMF and the World Bank) and more the outcome of specific social processes (Brenner 1986), which may or may not be reinforced by certain favourable – and equally contingent – circumstances within the global economy. The failure of structural adjustment is not simply the outcome of a bargaining process between international institutions and peripheral states over technical matters of policy, as some writers tend to imply (Mosley et al. 1991), but should also be seen as the outcome of a misguided attempt to impose an abstract model on concrete social relations.

The question of democracy

As already stated, part of the World Bank's agenda of "good governance" includes at least some attention to the need for the developing world to democratize. In the post-Cold War 1990s, aid has become increasingly tied to questions such as human rights issues and the creation of stable liberal democracies.

Given the West's historical record of support for some of the most authoritarian regimes in the Third World, and the widespread practice of using aid to boost markets for Western capital in the Third World, there are strong grounds for Third World suspicion of the new aid agenda. On the other hand, many of the criticisms of this agenda have come from those very same authoritarian regimes, who either fear for their future or may have actually been defeated through popular resistance and/or elections. In this respect, the West's promotion of "good governance" and democracy in the Third World cannot be dismissed out of hand.

However, there still remain strong grounds for questioning the West's commitment to democracy, and this again takes us back to the limitations of the neoliberal model, which basically informs the thinking of most Western governments. Just as states and markets are theorized (or more accurately fetishized) as ahistorical technical models, so too is the question of democracy. Liberal democracy is seen as the model for all societies, irrespective of specific histories and cultures. Even if the question of the appropriateness of the Westminster model is left aside (Munslow 1983, 1993), there are other questions that neoliberalism fails to address. Perhaps most important is the fact that neoliberalism itself has a very ambiguous relationship to democracy; indeed it treats democracy with the utmost suspicion. For neoliberals, democracy can exist in only a severely limited form. Rather than seeing democracy in the classical Athenian sense as a way of life, a specific culture based on widespread participation by citizens, neoliberalism sees democracy as simply a means to choose a government. Any major extension of the concept of democracy beyond the periodic election of government is viewed with suspicion by neoliberals because it threatens to undermine the all-important realm of freedom, which is guaranteed by a market society. According to neoliberals (Hayek 1960, Moss 1975), there is a distinction to be made between democracy and freedom, and the latter is given primary importance. The realm of freedom is guaranteed by the preservation of private property and a free market, which in turn guarantee independence from government, and therefore guarantee other "negative freedoms" such as the right to speak, associate and move about. The problem with democracy is that it can, if left unlimited, lead to demands being placed on government that undermine these freedoms. With an increase in the demands made on government, there is a corresponding increase in the amount of bureaucracy that is needed to deal with these demands, which in turn undermines even limited democracy. In the words of one former neoliberal Prime Minister:

> government has become more and more *remote* from the people. The present result of the democratic process has therefore been an increasing authoritarianism. (Thatcher 1986: 65)

Such a suspicion of democracy can be seen in the writings of the leading neoliberal writers on development issues. For instance, Deepak Lal (1983: 33) has expressed concern over the issue of vested interests in the developing world, and recommends that "[a] courageous, ruthless and perhaps undemocratic government is required to ride roughshod over these newly created special interest groups". Leaving aside the fact that this suggestion would require something more than a limited government, the argument of neoliberals appears to be that the preservation of the realm of freedom – that is, the market economy – takes precedence over the realm of democracy, the political sphere. In the *advanced* capitalist countries at least, too much government was caused by the electorate placing demands on democratically elected politicians, which led to the rise of public bureaucracy, which in turn led to restrictions on freedom and economic inefficiency. In the Third World, it is more difficult to blame democracy for too much government, and instead the problem is seen as one of corrupt dictatorship, which leads to "bad governance" (World Bank 1992). In this case, democracy is seen as part of the solution, but, in common with the neoliberal paradigm, it is the same, limited conception of democracy that is seen as providing the answer.

However, there is little sound basis to the assertion that democracy *per se* will guarantee economic growth (Healey & Robinson 1992: 94–112). As the previous chapter made clear, the most successful late developers in recent years have been authoritarian states in east Asia. Of course this observation does not mean that successful late developers need authoritarian states; instead, it again suggests that the problem of development is less a technocratic problem of governance and limited democracy, and more one of the social character of politics and the state. As Leftwich (1993: 620) argues, development "is not simply a managerial question, as the World Bank's literature on governance asserts, but a political one. For all processes of 'development' express crucially the central core of *politics*: conflict, negotiation and cooperation over the use, production and distribution of resources."

Moreover, the neoliberal suspicion of democracy leads one to ask a more fundamental question about both democracy and development. The neoliberal defence of private property and "negative freedoms" is fundamentally a social question and reflects a fear that participatory democracy might undermine the power of property owners in the context of highly unequal societies. There is a basic tension in neoliberal accounts of liberal democracy, as Beetham (1981: 191) makes clear:

> One can say that the equal citizenship implied in the idea of popular rule is restricted by the relationships of control and subordination constituted by ownership rights in the sphere of production, and the unequal weight these guarantee to a minority in the sphere of politics.

This statement is not meant to imply that there is no need for an individual citizen to enjoy personal "space" away from the collective (Beetham 1981: 199), but it is to argue that neoliberals constitute this boundary in such a way that unequal and unacceptable power relations are defended. In this way, unequal ownership of private property and other social inequalities can be said actually to *limit* and *undermine*, rather than guarantee, personal freedom.

These observations are all the more relevant in the context of many countries in the Third World, where "democratization" has gone hand in hand with neoliberal government policies that have increased social inequalities (see above). Moreover, one of the effects of the debt crisis has been to undermine the sovereignty of nation-states, at precisely the time when these same states are being encouraged to democratize. These factors not only serve to undermine democracy in its extended sense, but in the long run may even represent a threat to liberal democracy in its limited, neoliberal sense, as marginalized groups threaten rebellion and/or threatened elites rely on authoritarianism in order to defend their position (Leftwich 1993, Pinkney 1993: 165–7).

None of these negative observations is meant to imply that the process of democratization in the 1980s and 1990s is simply a charade. In many countries, significant gains, particularly in terms of civil rights, have been made. However, these are limited and their long-term future is far from guaranteed. What appeared to occur in the 1980s was a struggle over the ideal of democracy, which largely reflected the interests and agendas of different social groups within the Third World. Broadly speaking, elites within the Third World (as well as various Western agencies, politicians, and so on) espoused the neoliberal view of democracy, and aimed to keep democracy within strictly defined boundaries. On the other hand, there were struggles by social movements for a more participatory, egalitarian democracy

(Gills et al. 1993: 24–5). So, once again, democracy should be seen as a *social* process, rather than the technocratic one championed by neoliberalism.

Post-communism and development

Events around the recent collapse of communism in eastern Europe again clearly demonstrate the social nature of development, as opposed to the neoliberal technocratic approach. While some neoliberal writers regard the collapse as a sign of the "end of History" (Fukuyama 1992), whereby Western liberal democracy is said to be the highest goal for all of humanity, many neoliberals (not least in eastern Europe – see Naishul 1990) display a strikingly naïve faith in the ability of post-communist societies to become affluent, Western-style democracies. This section expands on the previous discussion by briefly examining three key issues: post-communism and theories of development; social relations in communist societies; and social relations and the development potential of post-communist societies. In so doing, the neoliberal conceptions of the market, the state and development are again challenged.

The collapse of communism and theories of development

The collapse of communism in eastern Europe has been accompanied by the growth of at least two implicit approaches to the development of post-communist societies. These two views broadly conform to the theories of underdevelopment and modernization discussed in Part I of this book, although in the case of the theory of modernization it now takes a more explicitly neoliberal position.

Neoliberals contend that the modernization of post-communist societies will inevitably occur if new governments adopt the correct policies of allowing market forces to flourish (see Fukuyama 1989). If governments leave their respective economies to competitive private enterprise, rather than the old rigid state monopolies of state communism, then this will lead to rapid economic growth on the basis of countries operating their comparative advantages. Neoliberal reformers in eastern Europe (Naishul 1990) place great emphasis on attracting foreign investment through the region's comparative advantage in cheap labour. In this way, reformers believe that "new Koreas" will develop in the region.

On the other hand, many writers on the left (Frank 1991) believe that the post-communist societies will simply function for the needs of Western capital, providing them with cheap labour, raw materials and labour-intensive manufactured goods. This school of thought therefore believes that the collapse of communism will lead to eastern Europe playing a subordinate role in the world economy. In this sense the Third World is said to be expanding rather than contracting.

As already discussed in Chapter 3, the problem with both of these views is that they provide us with an over simplistic, black and white dichotomy between inevitable rapid economic growth and inevitable economic stagnation. Contrary to the claims of neoliberalism, the rise of the NICs in east Asia is highly selective and hardly a product of the adoption of neoliberal policies; on the other hand, the rise of the NICs, no matter how selective, itself contradicts the claim that the Third World is doomed to a situation of eternal stagnation so long as it remains part of the world economy (see Chapter 5 for further details). Although it may be true that eastern Europe will suffer from competitive disadvantages in the world economy, as Western capital continues to monopolize the most advanced technology, skills, infrastructure and marketing (see also Chapter 5), these factors will impinge on different parts of eastern Europe in different ways. This suggests that there is again a need for an analysis of social relations within specific countries. Although a detailed analysis lies beyond the scope of this section, some cautious claims can be made. In order to undertake this task, I will first examine the social relations of production of state socialist societies, and then discuss the implications of these for capitalist development in post-communist societies.

Social relations in communist societies

A number of writers (Corrigan et al. 1978, Littler 1984) have emphasized how state socialist societies did not completely break with capitalist methods of work control. From as early as 1918, the Bolshevik leadership in the Soviet Union placed a lot of faith in hierarchical methods of work-place management. Under Stalin in the 1930s, a particular pattern of social control was established, by which the communist leadership undertook the task of developing the mass technology that had previously been developed in capitalist societies. For the Soviet leadership, as with other versions of orthodox Marxism (see Chapter 2), the primary task was seen to be the development of the productive forces, which was basically seen as mass industrial technology. From the late 1920s, the communist leadership also believed that the surplus for industrialization would be provided by the exploitation of the peasantry, through a process of "primitive socialist accumulation" (see Day 1977). In this way, state socialism was regarded by the communist leadership of the Soviet Union as the means to overtake Western capitalism. As a result, planners focused their attention on developing heavy industry, so that fewer resources were made available in light industry and agriculture. This "Soviet model" was to be repeated in post-war eastern Europe, and even in Maoist China (Bideleux 1985, White 1992).

However, this development strategy was undertaken in the context of social relations that were radically different from those of capitalist societies. Under capitalism, the separation of the producers from the means of production simultaneously means that commodity production is generalized, and so capitalists are compelled to cut costs and/or innovate in order to stay in business in a competitive environment (see Chapter 2). Such a situation did not exist in state socialist societies. Instead, the economy was subject to the will of central planners who were responsible for determining inputs and output targets in particular regions. In other words, plan targets would "filter down" to each particular enterprise, where specific plans had to be fulfilled. So, in state socialist systems, planning agencies, rather than individual firms, determined the accumulation process. Moreover, each particular enterprise did not risk going out of business if plan targets were not fulfilled, although managers risked removal from office if this problem occurred.

Furthermore, part of the inputs determined by central planners rather than by enterprise managers included labour. As full employment was guaranteed by the state, workers had a certain stake in the system, which was reinforced by other social rights such as a home and heavily subsidized basic food. Therefore, the successful implementation of hierarchical management strategies was severely restricted by the ability of workers successfully to resist management (as well as some shared interest in avoiding too difficult plan targets set by the centre – see Mandel 1989: 106).

Social relations in post-communist societies

As already stated above, neoliberals believe that the disappearance of the communist state and "correct" government policies will lead to the emergence of a rapidly developing capitalist society. However, once again this technical approach to development abstracts from the concrete social relations that exist in post-communist eastern Europe, as well as the hierarchies that exist within the global political economy. In terms of the latter question, the competitive advantages (economies of scale, more advanced infrastructure, skilled labour – see further Chapter 5) that "advanced" capitalist societies enjoy over the rest of the world are likely to ensure that most capital investment will continue to be concentrated in a few regions of the world, and that most of eastern Europe is unlikely to attract massive amounts of productive foreign investment.

In terms of the concrete social relations that exist in post-communist eastern Europe, it is far from clear that the region as a whole is moving unproblematically towards rapidly growing capitalist economies. It should be clear from the discussion in Chapter 2 that the emergence of capitalism depends on the outcome of contingent factors, rather than being the simple outcome of correct government policies. The attempted imposition of neoliberal government policy has had unexpected consequences, which have in some respects actually *strengthened* the command economies that existed under communism (Clarke 1993). The undermining of the communist state has strengthened local monopolies, which are no longer subject to control by the party-state apparatus, and in turn helped to maintain the negative power of workers on the factory floor (although this is not to deny that workers in some sectors have been defeated and are now jobless). Liberalization policies throughout eastern Europe have intensified this process, as structural adjustment has led to an inflation–debt spiral, which in turn has further encouraged the growth of the barter economy.

Essentially, then, the pre-existing system of production has maintained itself – although this is not to deny that defeats of the working class have taken place and unemployment has soared. Indeed, the growth of merchant capital activity (that is, trading rather than *productive capitalism*) has in some respects strengthened the old system of production, with devastating results for most of the population in eastern Europe. Burawoy (1992: 783) has accurately described what has occurred in Russia (which can be further generalized – on Poland, see Kowalik 1991: 275):

> The disintegration of the party state leads neither to chaos nor to successful economic reform ... but to an economy based on monopoly, barter and worker control. Rather than moving toward modern capitalism, the economy exaggerates pathologies of the old system. No longer restrained by the state, monopolies become stronger. As shortages become more severe and enterprises become more autonomous, barter becomes more important. As managers have to devote more attention to garnering supplies, as the party is no longer available as a tool of discipline, as enterprises can offer less to their labor force, worker control over production becomes stronger.

Although China may be said to have travelled further along the route towards a capitalist economy, this transition is also uncertain. Despite the very real concessions that have been made to private enterprise, the command economy of the Communist Party central planners remains intact. Whether or not this will remain the case is a question not of correct policies, but of the success or otherwise of political and social forces within China, and perhaps ultimately it will depend on whether the party-state bureaucracy can appropriate state property and then establish itself as a new bourgeois class (Smith 1993: 98). Once again then, the question of the transition to capitalism is a fundamentally *social* question, just as it was in England in the seventeenth century onwards.

Conclusion

This chapter has argued that neoliberalism reduces development to a purely technical problem, whereby states adopt correct policies of privatization and liberalization. Although in recent years international organizations influenced by neoliberal thinking have attempted to highlight the institutional framework in which policy is made, I have argued that the question of governance is still too technically oriented and abstracts from the social forces that ultimately determine development. The implementation of particular neoliberal policies has therefore had unforeseen consequences in many Third World – and now post-communist – countries, and, from the point of view of meeting the basic needs of the populace, such policies have often been disastrous.

7

The politics of the impasse II: challenging Third Worldism

This chapter examines what might immediately appear to be three relatively unconnected issues. These are: first, the changes in the global political economy that have undermined the unity of the Third World as a political force; secondly, the relationship between power relations in the global political economy and cases of Western intervention in the Third World; and thirdly, the relationship between Western conceptions of modernity and universalism, and relativist and "post-modern" challenges to these notions.

The chapter therefore examines each of these issues in separate sections. However, in attempting to maintain consistency with the first part of this book, the internal relationships between these issues are drawn out, partly by a close empirical examination of some of the issues, such as Western intervention. The chapter can be read as a summary of the philosophy of my whole argument, and is therefore deliberately (even) more polemical than the rest of the book. The first section therefore contextualizes the discussion by examining the reasons the Third World can be said to have declined and, as a precursor to the following sections, suggests some of the implications of this for an assessment of nationalism in the periphery. The second section then expands on these arguments by examining arguments for and against Western intervention in the periphery. Finally, the third section draws the argument together through a critical dialogue with post-modern accounts of the relationship between the West and the Rest.

The "end" of the Third World

The Third World emerged as a political and economic force in the 1950s, in the context of decolonization and the Cold War. The Non-Aligned Movement was founded in 1961, although its origins go back to the Bandung Conference of African and Asian countries in 1955. The basic principle of this organization was to apply a policy of non-alignment between the capitalist West and communist East (Willetts 1978: 18–19). In addition to this organization, the United Nations Conference on Trade and Development (UNCTAD) was formed in 1964, which focused its attention on the inequities of the international economic order, and particularly argued that the Third World suffered from unfavourable terms of trade in the global economy.

Third World solidarity reached a peak in the mid-1970s when, in the wake of the OPEC oil price rises of 1973–4, both UNCTAD and the Non-Aligned Movement called for a "new international economic order". These organizations used their effective majority in the United Nations' General Assembly to pass two motions in 1974–5, calling for the following:

(i) a new system of equitable trading that would guarantee compensation for the Third World for any shortfall in export earning;

(ii) a transfer of financial resources to the Third World;

(iii) the promotion of the transfer of technology from the First to the Third World;

(iv) regulation of transnational corporations (cited in Anell & Nygren 1980: 189–90).

In practice, few of these proposals were implemented in a world political economy where the institutions with real power all had (and still have) inbuilt majorities for the "advanced" capitalist countries (although this is not to deny the conflicts that take place within this group of countries – see Corbridge 1986: 212–15). Moreover, even when some reforms were made, such as the Lomé agreements between the European Communities and the ACP (African, Caribbean and Pacific) agricultural producers, they were limited in their scale, and, more relevant to my discussion here, they were made at the expense of other Third World producers (for instance in Latin America), rather than on the principle of collective Third World solidarity (Coote 1992: 122–9).

This point suggests that, for all the rhetoric of international solidarity, most developing countries followed the principle of promoting their own interests before those of "South–South" co-operation. Even the oil price rises of 1973–4, which at the time were regarded as an example of action that reflected a new balance of power in the world economy, can ultimately be seen in this light, because their effect was to alter the trading situation of oil exporters, but partly at the expense of the oil-importing countries in the Third World. Unity within the Non-Aligned Movement was similarly fragile, as in practice most of its members more or less sided with one or the other superpower in the Cold War (Worsley 1984: 325).

The high point of Third Worldism was at the time of the calls for a new international economic order in 1974–5, but by the early 1980s the fragility of this unity had been exposed. This was the case for at least four reasons (McGrew 1992, Kiely 1992b). First, as I have already implied, the unity of the Third World was always weak, and the changing economic and political climate in the 1980s intensified this fragility. The 1980s saw ideological conflicts such as the Iran–Iraq war, and increasing economic differentiation between Third World states. These processes made it increasingly difficult to identify a homogeneous bloc known as the Third World, a point rightly emphasized by those "post-impasse" writers that stress the increasing diversity of the so-called periphery (see Schuurman 1993: 29–32). Secondly, in the "advanced" capitalist countries, a number of governments were elected that were hostile to any notion of Third Worldism. This was exemplified by Reaganite conflation of unrest in the Third world with communist expansion, and US backing for terrorism throughout the world (although this was hardly new). The Thatcher–Reagan dismissal of the two Brandt reports (Brandt Commission 1980, 1983), which followed similar (though less confrontational) principles to those of the new international economic order, also showed a clear hostility towards any restructuring of the world system. Thirdly, recession in the 1980s had the worst effect in the South, which was reflected in declining GNP levels, the debt crisis and famine. For instance, in the period 1980–88, GDP declined in real terms by 14 per cent in Africa, 10 per cent in the Middle East, and 7 per cent in Latin America and the Caribbean (Freeman 1991: 154). The result of this unfavourable situation was that both individual nation-states and UNCTAD found themselves bargaining from a position of weakness in the global economy. Finally, the end of the 1980s saw the collapse of the Second World and the end of the Cold War. One consequence was that the notion of a world order divided into three worlds was out of date, and non-alignment lost much of its clarity.

Together, these factors contributed to the undermining of the idea of a Third World. However, it has also been suggested that the events of the 1980s simply intensified processes that were already visible, and that the concept of a united Third World was always something of a myth. This argument was suggested in Chapter 3, where it was argued that an over emphasis on the division of the world into nation-states abstracted from the conflicts that occurred *within* nations. This factor was a particularly acute one for proponents of a new international economic order, because the focus was on redistribution of wealth *between* nations. Therefore, there was no guarantee that a new international economic order would redress inequalities in wealth and income in the world economy – instead, a transfer of wealth from the First to Third World would just as likely lead to the impoverishment of the poor in the former along with increased wealth for the already wealthy in the latter. This critique was made by the right as well as the left (see Bauer 1981: 116–17), although, in the case of the former, a similar concern with *existing* levels of income and wealth inequalities is conspicuous by its absence (Sen 1984: 293).

Nevertheless, the point remains that conflict exists within the Third World, and this cannot simply be read off from the machinations of "Western imperialism". As Chapter 3 argued, to do so is to deny the capacity of the peoples in the periphery to forge their own history. It *is* the case that power is concentrated firmly in the hands of the Western powers, but it is *not* the case that this affects all nations of the periphery in a uniform way.

It is in this light that there is a basis for a reassessment of nationalism, and therefore of the case for intervention by Western powers, in the developing world.

Third World nationalism and Western intervention

There is a long history of Western intervention in the periphery, which can easily be denounced as imperialism. This applies to the colonial period and to the alarming number of interventions that have taken place since 1945. These interventions occurred for a variety of reasons, such as access to important raw materials, strategic interests in the context of the Cold War, and (not least) a US political culture in which the ruling elite has consistently believed that it has a divine right to expand beyond its territorial boundaries (Kiernan 1980). Western rhetoric concerning the promotion of democracy against communism during the Cold War can be dismissed as total nonsense when one considers the countless interventions designed to prop up right-wing dictators and even overturn liberal democracies (see Blum 1986, Pearce 1982). Moreover, during the Cold War, Western intervention in the Third World was far more common than so-called communist expansion (Halliday 1983: 97–104).

Nevertheless, there is now a belief in the West, even among those on the left, that there is a case for Western intervention in the Third World in the post-Cold War era. This is said to be because "oppressed peoples are looking for forms of western intervention that can save them from the horrors visited on them by their 'own' and neighbouring regimes . . . To uphold national sovereignty and damn intervention is to give a free hand to genocide" (Shaw 1993: 16). What is crucial here – at least from the point of view of the impasse in development studies – is that Shaw justifies intervention on the basis of the sound observation that conflicts exist *within* the Third World, and these cannot simply be read off from the actions of an omnipresent "West". This is made clear when Shaw (1993: 17) argues that "[t]he left has a particular duty to respond, not to the self-serving nationalist rhetoric of corrupt and repressive third world governments, but to the people who suffer from them". This statement echoes Warren's critique of (some versions of) dependency theory, which all too easily justified a reactionary nationalism in the name of so-called anti-imperialism (Warren 1980: chs 1 & 7).

On the other hand, many people on the Western left argue that intervention and imperialism amount to one and the same thing, and they cite the history of reactionary and bloody interventions by the Western world since 1945 (or earlier – Chomsky 1993 takes us back to 1492). On this basis, interventions in the 1990s in the Gulf and Somalia are regarded as imperialist in character (Pilger 1993: 10–11). There are, however, competing strands within this school of thought, which I allude to below.

The problem with these two views is that they tend to talk past each other. Although both approaches may appeal to the justice of their respective positions, it is seldom spelt out what is meant by this concept, a weakness intensified by the one-sided nature of both approaches. On the one hand, the interventionists appeal to justice and the rights of subjects (rather than states) in the periphery, but they tend to do so in isolation from the real world of international politics. On the other hand, opponents of intervention focus on *realpolitik* and the bloody history of Western intervention, but in so doing they tend to provide no clear grounds for *any* forms of intervention. These points can be illustrated by an examination of the competing positions in the Gulf War.

The interventionists argued that the United Nations' action to remove Iraqi forces from Kuwait was largely justifiable (Halliday 1991). The best criterion for what constitutes a just war can be found in the work of Michael Walzer (1977). He argues that war is justified when it is in response to an act of aggression by one state against the territorial integrity of another. In a new edition of this work, Walzer (1992: xi–xxiii) has argued that the Gulf War constituted a just war. This is so for the following reasons:

  (i) the Iraqi invasion of Kuwait in August 1990 was against the wishes of its citizens, and the rest of the population;
  (ii) the declared aims of the UN forces were to liberate Kuwait, and to ensure that Iraq would be incapable of further aggression;
  (iii) the UN forces did not go on to overthrow Saddam Hussein or to occupy Iraq, except to guarantee some safety for the Kurds after their unsuccessful uprising.

On the other hand, others have argued that United States' imperialism is so omnipotent that the only correct position was to support the Iraqi regime. The United Nations is simply a tool of US imperialism, and the USA's chief concern was economic (oil) and/or strategic (the preservation of Israel and Arab client regimes). Proponents of this view pointed out the double standards by which Iraq was condemned for its occupation of Kuwait, while there were no calls for "just wars" against Israel, Indonesia or, in the past, South Africa (Samara 1991: 265–6).

This point is more relevant than the interventionists would sometimes have us believe, as I show below. First, however, the pro-Iraq position needs further clarification. The key argument of this position is that Saddam Hussein represented a challenge to the status quo in the Middle East, where there were great discrepancies between the wealth of Arab states, and local "comprador" classes deposited their oil wealth in Western banks. In this respect, the Iraqi takeover of Kuwait represented a liberation for that country (Samara 1991: 260–1).

There are strong grounds for dismissing this position as every bit as opportunist as that of the worst hawks in successive US administrations. Saddam Hussein's nationalism can hardly be described as progressive – he was an old ally of the United States, in particular during the latter stages of the Iran–Iraq war, his treatment of Kurds within Iraq has been brutal and he has persistently attempted to control the cause of Palestinian national liberation (Halliday 1990: 73). Simply to assume that Saddam Hussein was now a progressive anti-imperialist because he had fallen out with his old allies is naïve at best, and at worst represents a mirror image of the US approach that "our enemy's enemy is our friend" (Elliott 1992: 11). Furthermore, Iraqi treatment of those living in Kuwait during the occupation can hardly be described as a "liberation"; rather, it was characterized by extremely repressive measures against the population. Moreover, to point to isolated examples of successful social programmes in Iraq (Gowan 1991) is hardly sufficient (and indeed is patronizing) to secure progressive credentials. Once again, Warren's point that anti-imperialist rhetoric is not necessarily progressive seems pertinent.

A less extreme anti-interventionist position was to not take sides in the war but at the same time not to call for action against the Iraqi regime. The basic justification for this view was that the international order was so unjust and exploitative that no one had the right to impose their will on anyone else. Of course this view abstracts from the fact that the Iraqi regime had done just that, and it becomes a call for lack of action. The logic of this view is that there can be no change for the better until the glorious day of worldwide socialism. Moreover, this view implicitly rests on the view criticized in the previous chapter that the capitalist state always unproblematically serves the functional needs of capital, and so actions by capitalist states are always seen as inherently "bad". According to this view, the West intervened in the Gulf because it suited its interests, but has been less willing to intervene in Bosnia because such a position also suits its interests. Although I think that there is a great deal of truth in this assertion, it takes things too far. Just because the West has no intrinsic interest in

149

intervention in Bosnia does not mean that we should simply leave it there (or, worse still, appeal to the Yugoslavian "class struggle" in a way that totally abstracts from the concrete conditions in the region), as many Marxists in the West imply (see Callinicos 1993). Instead, when there is a case for some form of intervention (as I believe there is in Bosnia) there should be criticism of Western governments precisely on the grounds that strategic or economic interests should not determine foreign policy (Magas 1993). The common assertion that these interests always win the day is to dismiss the struggle for alternatives from the outset. Similarly, just because intervention in one place may take imperialist forms (such as in Somalia in 1992–3) does not mean that the case against *any* form of intervention is established.

Standard Western left views (which I show below have much in common with post-modernism) can again be seen as based on an approach that contributes to, rather than transcends, the impasse. The structures of international capitalism are seen as so universally bad that there is no room for reform within this system. Struggle for reforms against this system is thereby discounted at the outset. We are therefore forced back to the logic of a Frankian "pessimism of the intellect, pessimism of the will", as discussed in Chapter 3. As Elliott (1992: 11) argues, this perspective "proffered an *abstract internationalism* whereby the cure for all remediable ills was postponed to an indefinite future".

These views then are based on perspectives that have contributed to the impasse in development studies. The pro-Iraq position is based on a patronizing Third Worldist–dependency approach in which all the ills of a country are blamed on the West, so that anti-Western positions are automatically progressive (see my brief discussion of Pol Pot in Chapter 3). The anti-sanctions position rests on a similarly misguided view that the "world-system" is so omnipresent and bad that the call for reforms within it is doomed to failure (again see Chapter 3).

Does this mean, then, that the interventionist view is correct? In terms of the Gulf War, I think not. In terms of interventions in other places at other times, the only answer that can be given is that it depends on the concrete circumstances (rather than by recourse to an omnipresent imperialism that is assumed always to win the day). On the question of the Gulf War, the pro-intervention position abstracts from the motives that guided US-led intervention. As already stated, there were enormous double standards in the decision to punish Saddam's invasion while other equally illegal occupations had not led to military action or even sanctions. It does seem odd that interventionists such as Fred Halliday and Norman Geras supported the US actions in the Gulf but made no call for similar action against South Africa, Israel or Indonesia (Cockburn 1991). According to this view, the USA intervened in the Gulf in order to maintain its hegemony in the region, and to help preserve regimes that had entered into an effective partnership with the West whereby these regimes deposited oil profits in the metropolitan countries in return for military protection (Stork & Lesch 1990, Bromley 1991, Brenner 1991: 134).

The standard interventionist reply to this point was that, although the motives of the USA were not necessarily guided by principles of justice, the outcome of the war – the removal of Iraq from Kuwait – was sufficient to justify the US action (Halliday 1991: 16, see also Halliday 1990: 73). However, the point is that the motives that guided the war cannot so easily be separated from its outcome, and these motives could be deciphered by an examination of the historical record of US intervention (Kiely 1992b). It is simply wishful thinking to argue that selectivity in intervention was irrelevant in the context of the Gulf War. As Elliott (1992: 11) argues:

> even were we to grant that the demonstrable hypocrisy of the UN Security Council (infinite patience with Israel, implacable belligerence towards Iraq) was irrelevant to the present crisis; even if, discounting motivation, the relevant criteria were the consequences of a military "resolution", was it nevertheless not the case that *the motives and intentions of the combatants offered a clue as to the likely consequences of any war?*

On this basis of combining an analysis of justice with *realpolitik*, the Gulf War may have been justified (and this is highly questionable) on the basis of *means*, but the *ends* strongly suggest that the anti-intervention position seems to be far more convincing. On the basis of an analysis of US imperial history, the consequences of the war were largely predictable. These outcomes can be summarized (see Elliott 1992: 13, *New Statesman* 1991) as follows:

(i) The restoration of the status quo in Kuwait (albeit with persecution of Palestinians);

(ii) The preservation of the Ba'athist regime in Iraq. Although it may be the case that the United States wished to remove Saddam Hussein from power, given the nature of the opposition (Shi'ite Muslims and communists) it is highly unlikely that the USA wanted to remove the regime (Gowan 1991: 50). Indeed, the USA actually sanctioned the suppression of the Islamic-led rebellion on the first day of the ceasefire.

(iii) Little progress on long-term solutions in the Middle East. (The progress, if any, on this front since the Gulf War can be attributed not to US victory in the Gulf, but to the compromises made by both the state of Israel and some of the leaders of the PLO. The only Western power that can take any credit is Norway, and certainly not the United States.)

(iv) The deaths of thousands of people in the region.

Therefore, principled opposition to the war was based not on an unthinking Third Worldism by which Saddam was championed as an anti-imperialist, but on a more pragmatic view of what constitutes a just war and how US *realpolitik* was so great that it militated against the conduct of a just war *from the outset* (see Hitchens 1993: 75–84).

What this discussion suggests, however, is that cases for intervention or non-intervention require far more careful consideration than is usually given. What is required is an analysis of how the case for a particular intervention (be it a war or a lesser form of intervention) can or cannot be justified. Although interventions cannot be dismissed as imperialist on an *a priori* basis, neither can they be justified on the basis of abstract principles of justice that are divorced from the real world of international politics. One of the most problematic arguments of the interventionists is their selectivity – for instance, Martin Shaw has cited a number of examples where he believes that there has been a case for intervention (Bosnia, Iraq, Cambodia, Somalia), but conveniently leaves out others (Israel, Indonesia, until recently South Africa) where a similar case could be made. Once again, my point is not to deny that there might be a case for intervention, but to express concern with the way in which the "trouble spots" cited have largely reflected Western foreign policy concerns.

Finally, as a precursor to the next section, two points require further elaboration. First, the argument that anti-Western positions are necessarily progressive has been called into question, a point expanded on in the next section. Saddam Hussein's invasion of Kuwait was in no way progressive and deserved to be condemned – albeit through a long-term policy of sanctions rather than US-led war. Secondly, the case against war and intervention must be made on the basis of *universal* justice (but applied to concrete international politics). One of the ironies of debate around the Gulf War was how one ostensibly progressive and anti-Western position was so incapable of formulating a coherent alternative to the US-led war. Starting from a supposedly progressive anti-Western position that rejects universal values such as justice and truth, Jean Baudrillard (1991), the high priest of post-modernism, simply denied that a war could take place, or indeed had taken place (see Norris 1992). It is to a discussion of post-modernism that I now turn.

Post-modernity, the "West" and the "Rest"

At the heart of post-modern social theory lies a critique of the universal pretensions of the Enlightenment. This view, which became increasingly influential in eighteenth-century Europe, argued that the world could be objectively analyzed on the basis of universal principles of truth, justice and reason (see Hamilton 1992). Enlightenment reason posited the view that there is a sharp separation between objective facts and subjective values and that the social world can be "captured" and analyzed "scientifically" just as much as the natural world.

Post-modern social theorists challenge this radical separation of fact and value, and argue that all social theories are inherently value laden. According to this view, there exists no objective science or truth, and all science rests on implicit values that cannot be easily refuted. Post-modernism is therefore defined as "an incredulity towards meta-narratives" (Lyotard 1984: xxiv). It therefore follows that all truth claims are simultaneously claims to power. As far as the Enlightenment is concerned, post-modernists argue that its claims to "know" universal justice were in fact claims that the "West" is inherently superior to the rest of the world – that is, the West was regarded as a model for other parts of the world to follow. As Stuart Hall (1992: 312) states:

> Enlightenment thinkers believed that there was one path to civilization and social development, and that *all* societies could be ranked or placed early or late, lower or higher, on the same scale.

When looked at in this way, Enlightenment views of the world can be seen as rationalizations for colonialism, aid with strings attached, and Western intervention in the "uncivilized" world. This is because the West sees itself as the leader of the world, and so must teach others the real meaning of "universal" justice. Within this Euro-centric discourse, the "rest" is represented as an incomprehensible "other". Edward Said (1978: 42) has written of Western representations of the Middle East (the term itself is a social construction) in which "the essence of Orientalism is the ineradicable distinction between Western superiority and Oriental inferiority". Such (mis)representations of difference as "backwardness" were an intrinsic part of colonial discourse, which was confined not only to the "Orient" but to the whole of the colonial world (see Asad 1973, Arens 1977, Hulme 1986).

Post-modern social theory therefore suggests that the search for universal standards of truth and justice should be abandoned. The world is in fact composed of a plurality of language games, local "truths" and discourses. According to Foucault (1980: 131), "[e]ach society has its regime of truth, its 'general politics' of truth; that is, the type of discourse which it accepts and makes function as true".

Such a relativist approach to the world has increasingly attracted the attention of theorists of contemporary "development". The post-modern critique of "enlightened imperialism" is undoubtedly very powerful, and so some writers have attempted to apply this to *all* versions of development (Marglin 1991, Sachs 1992). In doing so, much is made of the ethno-centric assumptions of modernization theory and of the problems of the modern Western world. For instance, Marglin (1991: 3) points to the problems of environmental destruction and waste, spiritual desolation, the predominance of meaningless, alienating work, and neglect of the elderly. The intellectual dominance of the West is said to derive from the political dominance of those who believe in its superiority, rather than from the model itself. According to this view, then, *all* development projects are relations of domination based on power and/or knowledge (DuBois 1991: 13).

Such relativist approaches therefore argue that practices in the Third World should be seen in their specific, culturally embedded context, since "there is no way of assessing their truth or falsity apart from people's beliefs" (Marglin 1991: 13). Therefore, practices such as female circumcision or arranged marriages should be seen in their correct context, and not subject to a critique from the supposedly enlightened West.

The great strength of this approach is its challenge to modernist hubris and the Euro-centric belief that the West is the model for others to follow. As argued in Chapters 2 and 3, such views ignore the relations of conflict that exist both within the Western world and between the West and the rest of the world (or, more accurately, between specific social groups in these regions). However, there are strong grounds for suggesting that these relativist approaches go too far.

First, as many critics have pointed out (see Habermas 1987, Dews 1987, Norris 1990), extreme versions of relativism are self-contradictory. To argue that there are no criteria by which different discourses can be assessed "looks itself suspiciously like an *absolute* claim to validity, and this is something which relativists hold to be impossible" (McLennan 1992: 339). Moreover, any understanding of another culture automatically leads to the acceptance that "some things can be asserted as meaningful across very different cultures" (ibid.). If this is the case, then it is at least possible that there are some universal criteria for the assessment of all societies.

Furthermore, without any universal criteria for the assessment of different societies, we are left with a situation where social science simply "rubs up against" the existing state of affairs. This is ironic given that relativism professes to be a "theory" that preaches tolerance and pluralism. In fact, at its worst it simply ignores, or even becomes an apology for, all kinds of oppressive practices.

For instance, some feminist critiques of modernity have correctly (although rather one-sidedly) emphasized that the Enlightenment project was in fact a highly particularistic one, based as it was on the emancipation of (white, bourgeois) men (Diamond & Quinby 1988: xv), and so its universalist pretensions in fact empowered only (some) men "at the expense of women and people of color". (I leave aside the totalizing implications of this statement, as well as its reliance on the crudest form of Frankian underdevelopment theory.) Feminists influenced by post-modernism have therefore drawn the conclusion that the Enlightenment project of emancipation by human agents should be abandoned because it imposes an artificial – and dangerous – unity on what "in fact" is a highly fragmented society (Butler 1990). The problem with such views is that resistance is reduced to private, highly individualized acts that fail to address the real issues of social power in modern societies; Butler (1990: 338), for example, champions the "resistance" of the drag performer who challenges essentialist accounts of gender identity and thereby exposes the "radical contingency in the relationship between sex and gender". As Nussbaum (1992: 212) argues, the effect of such an extreme representational anti-humanism is actually counterproductive, for "to give up on all evaluation and, in particular, on a normative account of the human being and human functioning [is] to turn things over to the free play of forces in a world situation in which the social forces affecting the lives of women, minorities, and the poor are rarely benign".

This criticism can equally be applied to the pragmatism of Rorty (1991: 258) and his followers who encourage feminists to embrace a philosophy that "gives up the claim to have right or reality on its side". Although this recommendation is accompanied by a number of suspiciously "realist" statements (for instance, slavery is *"absolutely wrong"* irrespective of time and place – ibid.: 258, my emphasis), Rorty also at least recognizes that his version of pragmatism is "as useful to fascists like Mussolini and conservatives like Oakeshott as it is to liberals like Dewey" (ibid.: 255). Once again, such relativist notions fail to challenge the realities of power in the world today. These can be effectively challenged (at least on an intellectual level) only by a *reconstruction* of the emancipatory ideals of the Enlightenment. In this respect, (some) feminists can picture feminism as a movement of resistance to a certain kind of *false pretension* to universality, namely "the appropriation by the male sex of an unfair share of natural goods and symbolic space" (Lovibond 1992: 72). Such a perspective does not reject "emancipatory meta-narratives"; instead, feminism's goal is rightly seen as "the liberation of women from all forms of domination, . . . [as] a thoroughly modern movement insofar as it claims these ideals as legitimate and necessary for women" (Hewitt 1993: 80).

These problems of relativism are similar when development issues are considered. This can be seen most clearly in the work of Marglin (1991), which, despite protestations to the contrary (1991: 26), actually ends up apologizing for oppression in "traditional" societies. For instance, one chapter in his volume argues that the introduction of smallpox vaccination to India by the British was an act of imperialist domination, because it led to the eradication of the cult of Sittala Devi, the goddess to whom one prayed to avert smallpox (Apffel Marglin 1991). The contributors to this work argue that this is another example of Western neglect of difference, and is based on the binary opposition between health and illness, and life and death (Marglin 1991: 8).

Such views are based on a crude interpretation of the work of Jacques Derrida, whose approach to deconstruction does not entail the view that representations of "reality" are completely devoid of referential content (Derrida 1989). Moreover, such a celebration of "tradition" essentializes cultures in such a way that conflict is simply written out of the picture. In this respect, cultural relativism actually shares a methodology that is close to functionalist sociology and/or philosophical utilitarianism. As regards the former, relativism tends toward the view that if something exists in a "traditional" society, then it must serve a function for that society – for example, female circumcision is examined not on the basis of the reality of gender *conflict*, but through a recognition of its culturally embedded context (which has no room for conflict) (Marglin 1991: 12–14). Moreover, in maintaining a strict divide between "traditional" and "modern" forms of knowledge (a case of a binary opposition that can be deconstructed in a Derridean way), relativism actually mirrors modernization theory, which is itself derived from functionalism (Long & Villareal 1993: 163). In terms of relativism's close association with utilitarianism, this can be seen most clearly in the work of Herrnstein-Smith (1988), who argues that no normative evaluation is better than another, and so only the market can decide what is effective in society. Although relativists would want to avoid such a political conclusion, it is clear that there is a close linkage between the rejection of any universal norms and the celebration of utility-maximizing individualism.

Indeed, the relativist's suspicion of the homogenizing thrust of universalism (the universal subsumes the particular) closely parallels the neoliberal-utilitarian suspicion of democracy (the democratic polity subsumes individual freedom), which was critically discussed in Chapter 6. Moreover, relativism actually goes further:

> [it] refuses to subject preferences, as formed in traditional societies, to any sort of critical scrutiny. It seems to assume that all criticism must be a form of imperialism, the imposition of an outsider's power on local ways. Nor does it simply claim (as do utilitarian economists) to avoid normative judgments altogether,

for it actually endorses the locally formed norms as good and even romanticizes them in no small degree. It confers a bogus air of legitimacy on these deeply embedded preferences by refusing to subject them to ethical scrutiny. (Nussbaum 1992: 232)

The relativist celebration of tradition can also be seen in its approaches to peasant agriculture and the environment. For instance, Marglin (1991: 8) argues that subsistence agriculture is preferable to commercial agriculture. I argued in Chapter 3 that the effects of commercialization are hardly beneficial to all, as the Green Revolution (to cite one of many examples) showed. However, this does not mean that subsistence agriculture in itself is any better, not least because it ignores the exploitative relations that existed in non-capitalist societies (see Bernstein 1990b: 69–72). Such romantic views apply with equal force to those versions of environmentalism that uncritically celebrate a "pre-modern" respect for nature (see for instance Goldsmith 1992: xvii). In fact, such views are fully compatible with those Western (mis)representations which patronize the "rest" as a romantic other (Said 1978: 118–19). The reality of the environment in pre-modern societies is in fact very different from what romanticists would have us believe. As Harvey (1993: 29) argues:

Faced with the ecological vulnerability often associated with such "proximity to nature", indigenous groups can transform both their practices and their views of nature with startling rapidity. Furthermore, even when armed with all kinds of cultural traditions and symbolic gestures that indicate deep respect for the spirituality in nature, they can engage in extensive ecosystemic transformations that undermine their ability to continue with a given mode of production. The Chinese may have ecologically sensitive traditions of Tao, Buddhism and Confucianism (traditions of thought which have played an important role in promoting an "ecological consciousness" in the West) but the historical geography of de-forestation, land degradation, river erosion and flooding in China contains not a few environmental events which would be regarded as catastrophes by modern-day standards.

Once again, these observations do not suggest that the "modern" world is better than the "traditional" world, but neither is it the case that the traditional world was an environmental Utopia. Again, Harvey (1993: 30) is useful on this point:

> The point here is not to argue that there is nothing new under the sun about the ecological disturbance generated by human activities, but to assess what exactly is new and unduly stressful, given the unprecedented scale of contemporary socio-ecological transformations.

So, although relativism starts out with the intention of preaching tolerance and the recognition of difference, it comes dangerously close to a celebration of repression. In this way, it echoes those naïve views of the left that uncritically celebrated Saddam Hussein's invasion of Kuwait, and similarly champion – or at least refuse to criticize – other reactionary nationalist leaders simply because they are "anti-West". For instance, many post-modern writers refused to criticize human rights abuses in Iran after the revolution. Thus, according to Baudrillard (cited in Bruckner 1986: 181–2):

> It makes no difference if it is at the cost of religious "fanaticism" or moral "terrorism" of a medieval sort. For better or for worse, it is undeniable that a ritual viciousness, one that is not at all outdated, a tribalism that does not accept Western models of a free society, can pose a real challenge to such a world order.

What is startling about this perspective is its complete ignorance of the dynamics of the Iranian revolution, as well as its philosophy of "my enemy's enemy is my friend". It is also interesting how Baudrillard (and, it must be said, many others on the Western left) was so prepared to take the anti-imperialist rhetoric of the Iranian leadership at face value, something that subsequent events (the Iran–Contra scandal) showed to be hollow.[1]

Moreover, the political dangers of relativism go further than this. Again it needs stressing that relativism provides no means for assessing truth. But if this is the case, then "[w]hat possible ground could we have, on these assumptions, for preferring the testimony of death camp survivors or the archival research of serious historians of the Holocaust over the accounts of revisionists who deny Nazi genocide? Why think *pro-apartheid* apologetics qualitatively different from or inferior to the work of liberal, radical or socialist South African writers?" (Howe 1992: 37). A few more examples show the absurdity of the need to ask such questions. If we accept the argument of the Marglins that the "binary opposition" between life and death is a Western one, then we have no way of effectively criticizing the massacres carried out by the colonial powers. Neither do we have any grounds for complaining when food is exported from one impoverished region or country to another, richer area, because starvation is a concept based on the Western binary opposition between life and death. Then there is the aforementioned contention of Baudrillard (see Norris 1992: 192–6) that the Gulf War did not take place. Need I really go on citing such absurdities?

One more example is important though, because it shows how relativism actually mirrors those (right-wing) versions of the Enlightenment that posit the view that "West is best". This is the case of Western reactions to the Rushdie affair. In this case some writers (among them Anthony Burgess and Fay Weldon) argued that the fatwa against Rushdie showed that the West had a superior culture in that it preached tolerance, as against an intolerant and oppressive religion (Islam). Some on the left (at least in the West – the left in the Middle East was generally very supportive of Rushdie) were less than willing to support Rushdie, and this was justified on the basis of a respect for other cultures. However, what *both* sides failed to do was challenge the notion of Islam that was cited in these debates. Both sides shared an essentialist view of Islam, which reduced it to an ahistorical religion devoid of any conflicting interpretations. In fact, "Islamic fundamentalism" refers to a variety of movements that are highly *political* and *modern* in character, and which are based on only one of a number of interpretations of the Koran (see Al-Azmeh 1993: 10–14, Hitchens 1993: 289–302). In this respect, "fundamentalism" itself dehistoricizes Islam, a position shared by Euro-centrics and relativists alike (Al-Azmeh 1993: chs 4 & 7). The convergence between these supposedly diverse positions shows how politically correct "anti-colonial" relativism actually "reproduce[s] all the worst features of colonialist writing itself; of the myths of group destiny and cultural exclusivism fostered by the propagandists of imperial expansion" – in this way, such views have much in common with those of Pol Pot and Slobodan Milosevic (Howe 1992: 37, see also Hughes 1993).

Of course many relativists would halt at some point and would criticize different forms of oppression. For instance, Lyotard (1984: 66–7) has attempted to rescue the concept of justice from a complete relativism, and Rorty's comments on slavery cited above suggest that his pragmatism is compromised by at least some commitment to universalism. But, if this is the case, then how consistent are they being when the starting point of post-modernist accounts is a rejection of universal principles? Without some acceptance of universalism, how can different forms of oppression be criticized? Post-Marxists such as Laclau (1990) have attempted to avoid this trap by stressing the need (a universal one?) for different emancipations based on radical democracy. However, Laclau deliberately avoids any description of what actually constitutes radical democracy, on the grounds that "ambiguity and indeterminacy are central features of democracy" (ibid.: 169). The problem with this view is of course that democracy then becomes simply how a particular locality chooses to define it.

What is useful in the work of the relativists is the critique of those Western notions of justice and truth that are used to defend oppressive practices. As should be clear from other chapters in this book, one of the worst problems in development studies has been to impose abstract *models* of development in very different social contexts, and so these models become forms of imposing power on societies. This is undoubtedly the great strength of post-modernist critiques of social theory – although it should be noted that this critique is hardly a new one and is closely associated with the work of Karl Popper (1986). Neither is it limited to writers who associate themselves with an "anti-truth" position (see Said 1978, 1993a, 1993b: 28, also see the "soft post-modern" accounts, which still hold on to some notion of universal value – Slater 1992, Parpart 1993, Squires 1993). Indeed, the critique is at least as old as the Enlightenment itself (see Hamilton 1992). What is unacceptable about "hard" post-modern critiques is the tendency to lapse into nihilism. In other words, it is precisely because these models have led to oppressive practices that we need a better, more truthful account of the world in order to challenge these practices. In this respect, deconstruction is best seen as an expression of an attitude towards truth of "never simply" rather than "simply never" (the terms are taken from Norris 1993).

Of course, the search for better accounts of the world is ultimately a *practical* activity; that is, an activity that can never be divorced from the concrete situation of "real, living individuals". However, social science can still play a role in at least providing a limited framework for specifying the conditions for a better life for humanity. This problem of course is precisely what pragmatists like Rorty and relativists like Lyotard reject, but in so doing they fall back into a political position that is potentially a crude justification for the way things are.

What is therefore needed is an alternative position – one beyond both Euro-centrism and relativism. As already argued in Part I of the book, a critical Marxism is useful as an account of the different forms of capitalist development in the modern world. However, in terms of the search for an emancipatory discourse, what is also needed is an account of what constitutes *universal human needs*. This is especially so in development studies, where the ethical assumptions of conflicting positions are rarely clarified.

The basis for such a universalism is the recognition of others as human, regardless of place of origin or habitation, and a basically shared consensus about features whose absence would mean the end of human life (Nussbaum 1992: 216–20). Put differently, the search for universal principles such as needs and rights begins with the recognition that there are certain features that *all* human beings have in common, regardless of specific culture. Such needs would include the need for food, clothing and shelter, sexual desire, the capacity to think, and the capacity for pleasure and pain. Any society should, as a minimum, secure these basic needs for all of its citizens. This view is not Euro-centric because Western societies should equally be assessed on their success or failure in meeting these criteria, and, moreover, it leaves open the question of why these basic needs may not be met in any particular society – and it is at this point that a universalist critique of Western domination of the global economy is important. Moreover, this view is not imposing an artificial unity on the citizens of any particular society; instead, it is laying down the conditions for active participation by all citizens, *so that plurality may then be secured*. Thus, for Doyal & Gough (1991: 160), basic needs such as health and autonomy are preconditions for further activity so that individuals in specific cultures can "choose the activities in which they will take part within their culture". Habermas (1992: 240) has similarly recognized that a genuine universalism does not preclude the recognition of difference:

162

What does universalism mean, after all? That one relativizes one's own way of life with regard to the legitimate claims of other forms of life, that one grants the strangers and the others, with all their idiosyncrasies and incomprehensibilities, the same rights as oneself, that one does not simply exclude that which deviates from it, that the areas must become infinitely broader than they are today – moral universalism means all these things.

This genuine recognition of difference among citizens can be contrasted with utilitarian approaches to difference, which fail to condemn extremes of wealth and poverty, or overconsumption and hunger. Given my comments about relativism above, along with the view of some writers that the "binary opposition" between life and death is a Western conception (and so, one assumes, this perspective must also be applied to famine), then they too must be put into the same camp as the utilitarians (see Soper 1993: 115, and more generally the "universalist" work of Sen, especially Sen 1981, 1984).

Finally, these rather abstract points are best illustrated by a discussion of one contemporary and more concrete example, which shows clearly how both Euro-centrism and relativism must be transcended – that of human rights. The 1993 World Conference on Human Rights saw two clear positions that illustrate my argument. The first view, which is closer to the relativist position, was taken by some governments (such as China and Indonesia) in the Third World. This view holds that it is a form of cultural imperialism for the West to impose its own (culturally specific) view of rights on the rest of the world (see Ching 1993: 27). The Western view is that there are certain universal rights that should be applied to all countries. The problem with this view is that, even if we leave aside Western abuses of human rights in the First and Third World (one thinks again of imperialist interventions discussed above), the actual content of these rights is limited at best. For instance, Western conceptions have very little to say about the social and economic inequalities of the global order (Rajamoorthy 1993) – rights appear to be limited to those associated with a liberal democratic political order. But, as some of the relativists on human rights have argued, what is needed is some attention to economic rights and the "right to development" (Ching 1993). According to this view, the West is on the one hand preaching to the South about the need for human rights issues to be addressed, but failing to address how issues such as debt repayments and global inequalities are themselves human rights abuses. (Note that these relativists are very different from those relativists discussed above, who reject development as a Western conception. However, both approaches are applying relativist principles, albeit to different subjects.)

There is undoubtedly much truth in this argument. However, there is a danger in taking things too far, so that a Third Worldist position once again becomes an apology for state oppression. While it may be hypocritical of Western governments to lecture Third World governments on human rights, so too is it hypocritical of governments of the South to talk about human rights when they are suppressing their own populations. Indeed, when governments talk of the cultural specificity of rights, they are often simply trying to justify repression against a population that does not share its views (for instance, on Korea see Berry & Kiely 1993). In this way, human rights abuses are ignored in the "struggle for development" – and "development" therefore becomes the search for economic growth for the benefit of a minority.

What these two extremes show once again is the need for a middle ground between Euro-centric views and Third Worldist views. As Brazier (1993: 6) argues, "we should not have to choose between the two different kinds of human rights, between freedom from want and freedom to express. We should demand both – not just for ourselves but for all the world's citizens." (For a similar universalist position see Amnesty International 1993.)

This discussion has not tried to imply that recognition of the universality of these rights will lead to their being more easily achieved. In much of the Third World the state is highly repressive and is hardly likely to stop being so just because of a universal commitment to human rights. An abstract commitment to universal rights is not the same as their realization. We therefore also need to explain concretely why certain organizations are such great obstacles to the realization of these rights. For example, historical-sociological analysis is needed to understand why many states in the Third World are so repressive. Such an analysis lies beyond the scope of this work, but there is I think the basis for comparative and historical work that examines the relationship between crime, state repression and the "primitive accumulation" of capital (see the suggestive comments of Linebaugh in Linebaugh 1993: xv–xxvii). It is precisely this kind of challenging work that would help the sociology of development to transcend its impasse.

However, the recognition of universal rights *is* important, irrespective of the claims of relativists or even ultra-left Marxists. It is only by recognizing the universality of rights and needs that we will be in a better position to help them to be fulfilled. *Mere* moralizing is certainly not the answer, but neither is the view that moral questions are merely Western or bourgeois (see Geras' useful comments on the ambiguities in Marx's work, in Geras 1985: 84–5). Indeed, the case for a more equitable global order, for freedom from hunger, and indeed for socialism, is based on highly moral principles.

Conclusion

This chapter has outlined three principal arguments: first that recent changes in the global political economy have undermined long-held assumptions that the world can easily be divided into a homogeneous North and unitary South; secondly, that this observation has at least enabled us to rethink the question of what is progressive about certain types of Third World nationalism, and in particular to challenge the view that "anti-Western" views are inherently progressive; and thirdly, that one should uphold the claims for genuine universal principles rather than uncritically celebrate a variety of "particularisms". In so doing, I have especially challenged the view that *all* forms of intervention in the Third World are imperialist. It is only those reactionaries in the Third World who support capital punishment who claim that Amnesty International is an imperialist organization. To repeat myself, none of these comments is meant to imply that the West can simply intervene when and where it likes, or that it is the rightful leader of the world. My defence of universal principles can be used to criticize Western imperialism in all its forms as much as human rights abuses by Third World governments (often backed by Western ones). What I am challenging is a spurious relativism that romanticizes the Third World and fails to see oppressive practices by ruling classes, privileged gender or ethnic groups, or state dictators. Such a relativism is a fantasy that is rooted in the very imperialism that it purports to criticize. As Nussbaum (1992: 240) states, "[t]he moral sentiments of this sort of relativism . . . are the sentiments of the tourist; wonder, curiosity, and amused interest".

Note

1. This did not, of course, stop some people on the Western left from giving their support to the Iranian leadership during the Rushdie affair (see text for details). This support was given on the most vacuous of anti-imperialist sentiments, which again took the rhetoric of the leadership at face value, *after* details of the Iran–Contra affair had been revealed. The logic of such a position was that Rushdie – who had written a book – was a lackey of Western imperialism, while the Iranian revolutionary leadership – which had undertaken arms for hostages deals with well-known imperialist thugs such as Oliver North – really was anti-imperialist.

8

Conclusion

The purpose of this book has been to identify the impasse in development studies and to suggest some of the ways that it might be transcended. In this respect the book can be seen as part of a growing body of "post-impasse" literature (see the collections edited by Schuurman 1993 and Booth 1994). Although I share the view that the impasse has arisen because of a commitment to theory that is too deterministic and dogmatic, I have taken issue with a number of writers (Booth 1985, 1994) who place the blame for the impasse on a commitment to Marxism.

Indeed, Part I of the book argued that development theory has all too easily constructed models of development and underdevelopment that are divorced from the real history and struggles of agents in the global political economy. Such *a priori* theories have therefore become divorced from the real world, and in doing so have *fetishized* (in Marx's sense) social reality. The impasse is therefore best transcended by closer attention to agency in the global order – the Marxist focus on *struggle*, rather than the fetishization of structures.

In this respect Marxism remains a very useful tool of analysis for understanding the different forms of capitalist development in the world today. Capitalist development is an open-ended, uncertain process that is ultimately determined by the struggles of concrete social actors. Studies that focus on the "real history of production relations" can therefore illustrate the diversity of concrete experiences of capitalist development. At the same time, the uncertainty of these struggles should not themselves be fetishized so that all we are left with is a post-modern "celebration of contingency", because this too evades the question of the real relationships of power that exist both within nation-states and in the world system.

Class therefore must remain a central category in development studies, because it enables us to understand the question of power in the world today *and* the struggles against those relations. In other words, it unites an analysis of both the power relations and the uncertainty of concrete cases of capitalist development. In this way, we are left in a stronger position to understand that development is not a simple question of technical policy, such as "getting prices right", but is ultimately a *social process*, as my chapter on the rise of the east Asian NICs made clear. Hence, neoliberalism, with its exclusive focus on questions of technical policy and its ahistorical approach to states and markets, also fetishizes social reality.

These comments are not meant to imply that everything is reducible to "class", or that gender or ethnic relations are simply epiphenomena of "the class struggle". Class remains an indispensable category for understanding the phenomenal forms of "actually existing capitalisms", but it is not the only basis of conflict in the world today (Larrain 1994: 13–18). In examining the real history of production relations, proper attention must also be paid to how social actors view the world and how this understanding is affected by socially constructed relationships of power. Therefore the study of actually existing capitalisms may involve concrete analysis of caste, gender and ethnic relations, as well as those of class.

Thus the impasse must be transcended by closer attention to social relations and the struggles that are inseparable from these relations. However, closer attention should also be paid to the *politics* of the impasse. Development studies has been bogged down by a dogmatism that divorces theoretical construction from everyday political questions, and in particular the rapid changes that have emerged in the global political economy over the past 20 years. The result has been a tendency – most marked among sociologists – to ignore the neoliberal counterrevolution. Chapters 5 and 6 argued that, in reducing development to a simple matter of policy-making, this particular ideology also fetishized social reality, and as a result is excessively optimistic about the development potential of much of the Third World and the post-communist societies. In practice, structural adjustment has all too often had unpredictable results (at least for those implementing the policies) and led to sacrifice (for many) without any long-term benefits. On the whole, neoliberal policies are hurting but they are not working. The failure of neoliberal policies must therefore call into question any notion that History has ended.

Finally, and perhaps most importantly, this book has defended the discipline of development studies. Although I have recognized that an impasse has existed for some time, it remains the case that the study of development must remain both critical and internationalist. Unfortunately, far too many social scientists now appear to equate criticism and internationalism with Euro-centrism. A great deal of development theory has suffered from this problem, but a retreat into a nihilistic relativism does not represent a viable alternative; indeed, it all too easily becomes an apology for oppression, as recent political events such as the Gulf War illustrate. Development studies does need to pay more attention to the specificity of particular localities, but this must not be at the expense of the recognition that there are universal needs, and that the struggle for these needs continues to lead to social conflict. Without such a recognition, development studies will undermine its own future and patronize the so-called Third World rather than critically engage with it.

References

Aglietta, M. 1987. *A theory of capitalist regulation.* London: Verso.

Alavi, H. 1989. Formation of the social structure of South Asia under the impact of colonialism. In *South Asia*, H. Alavi & J. Harriss (eds), 5–19. London: Macmillan.

Alavi, H. & T. Shanin (eds) 1982. *Introduction to the sociology of developing societies.* London: Macmillan.

Al-Azmeh, A. 1993. *Islams and modernities.* London: Verso.

Allen, J. 1992. Post-industrialism and post-Fordism. In *Modernity and its futures*, S. Hall, D. Held, T. McGrew (eds), 169–220. Cambridge: Polity.

Allen, T. 1992. Prospects and dilemmas for industrializing nations. In *Poverty and development in the 1990s.* T. Allen & A. Thomad (eds), 379–90. Oxford: Oxford University Press.

Almond, G. & J. Coleman (eds) 1960. *The politics of the developing areas.* Princeton, NJ: Princeton University Press.

Althusser, L. & E. Balibar 1979. *Reading capital.* London: Verso.

Amin, A. 1991. Flexible specialization and small firms in Italy: myths and reality. See Pollert (1991), 119–37.

Amin, A. & K. Robbins 1990. The re-emergence of regional economies? The mythical geography of flexible accumulation. *Environment and Planning D: Society and Space* 8, 7–34.

Amin, S. 1976. *Unequal development.* Hassocks: Harvester.

Amin, S., G. Arrighi, A. G. Frank, I. Wallerstein 1982. *Dynamics of global crisis.* New York: Monthly Review Press.

Amnesty International 1993. *World conference on human rights - oral intervention.* Bangkok: Regional meeting for Asia.

Amsden, A. 1985. The state and Taiwan's economic development. In *Bringing the state back in*, P. Evans, D. Rueschmeyer, T. Skocpol (eds), 78–106. Cambridge: Cambridge University Press.

——1989. *Asia's next giant.* Oxford: Oxford University Press.

——1990. Third world industrialization: "global Fordism" or a new model?

New Left Review **182**, 5–31.

Anderson, P. 1974. *Lineages of the absolutist state*. London: New Left Books.

Anell, L. & B. Nygren 1980. *The developing countries and the world economic order*. London: Methuen.

Apffel Marglin, F. 1991. Smallpox in two systems of knowledge. See Marglin (1991), 102–44.

Apter, D. 1965. *The politics of modernisation*. Chicago: University of Chicago Press.

Archer, M. 1991. Sociology for one world: unity and diversity. *International Sociology* **6**, 131–47.

Arens, W. 1977. *The man eating myth*. Oxford: Oxford University Press.

Asad, Talal (ed.) 1973. *Anthropology and the colonial encounter*. London: Ithaca Press.

Balassa, B. & J. Williamson 1987. *Adjusting to success: balance of payments policy in the East Asian NICs*. Washington, DC: Institute of International Economics.

Balassa, B., G. Bueno, P. Kuczynski, M. Simonsen 1986. *Toward renewed economic growth in Latin America*. Washington, DC: Institute of International Economics.

Baran, P. 1957. *The political economy of growth*. New York: Monthly Review Press.

Barnett, R. & R. Muller 1974. *Global reach*. New York: Simon & Schuster.

Barone, C. 1984. Reply to Hart-Landsberg. *Review of Radical Political Economics* **16**, 194–7.

Baudrillard, J. 1991. The reality gulf. *Guardian*, 11 January.

Bauer, P. 1981. *Equality, the third world and economic delusion*. Cambridge, Mass.: Harvard University Press.

——1984a. *Reality and rhetoric*. London: Weidenfeld & Nicholson.

——1984b. Remembrance of studies past: retracing first steps. In *Pioneers in development*, G. Meier & D. Seers (eds), 27–43. Oxford: Oxford University Press.

Beetham, D. 1981. Beyond Liberal Democracy. *The Socialist Register*, 190–206. London: Merlin.

Bello, W. & S. Rosenfeld 1992. *Dragons in distress*. London: Penguin.

Benton, T. 1986. *The rise and fall of structural Marxism*. New York: St. Martin's Press.

Berger, P. 1985. Speaking to the Third World. In *Speaking to the Third World*, P. Berger & M. Novack, 1–12. Washington, DC: Institute for Public Policy Research.

——1986. *The capitalist revolution: fifty propositions about prosperity, equality and liberty*. New York: Basic Books.

Bergesen, A. 1990. Turning world system theory on its head. In *Global cul-*

ture, M. Featherstone (ed.), 67–81. London: Sage.

Bernstein, H. 1971. Modernisation theory and the sociological study of development. *Journal of Development Studies* 7, 141–60.

——1979. Sociology of development versus sociology of underdevelopment? In *Development theory: four critical essays*, D. Lehmann (ed.), 77–106. London: Frank Cass.

——1982. Industrialization, development and dependence. See Alavi & Shanin (1982), 218–35.

——. 1987. Review article: John Sender and Sheila Smith – the development of capitalism in Africa. *Capital and Class* 33, 168–72.

——1990a. Agricultural "modernization" and the era of structural adjustment: observations on sub-Saharan Africa. *Journal of Peasant Studies* 18, 3–35.

——1990b. Taking the part of peasants? See Bernstein et al. (1990), 69–79.

Bernstein, H. & H. Nicholas 1983. Pessimism of the intellect, pessimism of the will? A response to Gunder Frank. *Development and Change* 14, 609–24.

Bernstein, H., B. Crow, M. Mackintosh, C. Martin (eds) 1990. *The food question*. London: Earthscan.

Berry, S. & R. Kiely 1993. Is there a future for Korean democracy? *Parliamentary Affairs* 46, 594–604.

Bettelheim, C. 1972. Appendix I: theoretical comments. See Emmanuel (1972), 271–322.

Bideleux, R. 1985. *Communism and development*. London: Methuen.

Blum, W. 1986. *The CIA: a forgotten history*. London: Zed.

Bonefeld, W. 1987. Reformulation of state theory. *Capital and Class* 33, 96–127.

Booth, D. 1985. Marxism and development sociology: interpreting the impasse. *World Development* 13, 761–87.

——1993. Development research: from impasse to a new agenda. See Schuurman (1993), 49–76.

Booth, D. (ed.) 1994. *Rethinking social development*. London: Longman.

Bozzoli, B. 1983. Marxism, feminism and South African studies. *Journal of Southern African Studies* 9, 139–71.

Bradby, B. 1975. The destruction of natural economy. *Economy and Society* 4, 125–61.

Brandt Commission 1980. *North–South: a programme for survival*. London: Pan.

——1983. *Common crisis*. London: Pan.

Brass, T. 1988. Slavery now: unfree labour and capitalism. *Slavery and Abolition* 9, 183–97.

Brazier, C. 1993. Waking up the world. *New Internationalist* 244, 4–7.

Brenner, R. 1977. The origins of capitalist development: a critique of "neo-

Smithian" Marxism. *New Left Review* **104**, 25–92.

Brenner, R. 1986. The social basis of economic development. In *Analytical Marxism*, J. Roemer (ed.), 23–53. Cambridge: Cambridge University Press.

——1991. Why is the United States at war with Iraq? *New Left Review* **185**, 122–37.

Bresheeth, H. & N. Yuval-Davis (eds) 1991. *The Gulf War and the new world order*. London: Zed.

Brett, E. A. 1983. *International money and capitalist crisis*. London: Macmillan.

——1985. *The world economy since the war*. London: Macmillan.

——1988. States, markets and private power. In *Privatisation in Less Developed Countries*, P. Cook & C. Kirkpatrick (eds), 47–67. London: Harvester Wheatsheaf.

——1988. State power and economic inefficiency: explaining political failure in Africa. IDS Bulletin 17(1), 22–9.

Brewer, A. 1980. *Marxist theories of imperialism*. London: Routledge.

Bromley, S. 1991. Crisis in the Gulf. *Capital and Class* **44**, 7–14.

Browett, J. 1985. The newly industrialising countries and radical theories of development. *World Development* **13**, 789–803.

Bruckner, P. 1986. *The tears of the white man: compassion as contempt*. London: The Free Press.

Bundy, C. 1979. *The rise and fall of the South African peasantry*. London: Heinemann.

Burawoy, M. 1976. The functions and reproduction of migrant labour: comparative material from Southern Africa and the United States. *American Journal of Sociology* **81**, 1050–87.

——1992. The end of sovietology and the renaissance of modernisation theory. *Contemporary Sociology* **21**, 774–85.

Burawoy, M. & P. Krotov 1993. The economic basis of Russia's political crisis. *New Left Review* **198**, 49–69.

Burnham, P. 1993. Beyond states and markets in international relations: Marx's method of political economy. Mimeo.

Butler, J. 1990. *Gender trouble: feminism and the subversion of identity*. London: Routledge.

Byres, T. 1989. Agrarian structure, the new technology and class action in India. In *South Asia*, H. Alavi & J. Harriss (eds), 46–58. London: Macmillan.

——1991. The agrarian question and differing forms of capitalist agrarian transition: an essay with reference to Asia. In *Rural transformation in Asia*, J. Breman & S. Mundle (eds), 5–72. Oxford: Oxford University Press.

Byres, T. & B. Crow 1988. New technology and new masters for the Indian countryside. In *Survival and change in the Third World*, B. Crow & M.

Thorpe (eds), 163–81. Cambridge: Polity.

Callinicos, A. 1993. Intervention – disease or cure? *Socialist Review* **165**, 7–9.

Caporaso, J. 1982. Industrialization in the periphery: the evolving global division of labour. *International Studies Quarterly* **25**, 247–84.

Cardoso, F. 1982. Dependency and development in Latin America. See Alavi & Shanin (1982), 112–27.

——1987. The dependency perspective. In *Latin America*, E. Archetti, P. Cammack, B. Roberts (eds), 13–18. London: Macmillan.

Cardoso, F. & E. Faletto 1979. *Dependency and development in Latin America*. Berkeley: University of California Press.

Chandra, R. 1992. *Industrialization and development in the third world*. London: Routledge.

Ching, F. 1993. Asian view of human rights is beginning to take shape. *Far Eastern Economic Review* 29 April.

Chomsky, N. 1993. *Year 501*. London: Verso.

Clarke, S. 1977. Marxism, sociology and Poulantzas' theory of the state. *Capital and Class* **2**, 1–31.

Clarke, S. (ed.) 1980. *One dimensional Marxism*. London: Allison & Busby.

Clarke, S. 1988. Overaccumulation, class struggle and the regulation approach. *Capital and Class* **36**, 59–92.

——1993. Privatisation and the development of capitalism in Russia. In *What about the workers?* S. Clarke, P. Fairbrother, M. Burawoy, P. Krotov (eds), 199–241. London: Verso.

Cliffe, L. 1982. Class formation as an "articulation" process: East African cases. See Alavi & Shanin (1982), 262–78.

Cline, William 1982. Can the East Asian model of development be generalized? *World Development* **10**(2), 41–50.

Cockburn, A. 1991. The war goes on. *New Statesman* 5 April, 14–15.

Cohen, G. 1978. *Karl Marx's theory of history: a defence*. Cambridge: Cambridge University Press.

Cohen, R. 1987. *The new helots*. Aldershot: Gower.

Colclough, C. & J. Manor (eds) 1993. *States or markets?* London: Clarendon.

Colletti, L. 1972. *From Rousseau to Lenin*. London: New Left Books.

Connell, R. W. 1984. Class formation on a world scale. In *For a new labour internationalism*, P. Waterman (ed.), 176–209. The Hague: International Labour Education Research and Information Foundation.

Coote, B. 1992. *The trade trap*. London: Oxfam.

Corbridge, S. 1986. *Capitalist world development*. London: Macmillan.

——1989. Marxism, post-Marxism, and the geography of development. In *New models in geography*, vol. 1, R. Peet & N. Thrift (eds), 224–53. London: Unwin Hyman.

——1990. Post-Marxism and development studies: beyond the impasse. *World Development* **18**, 623–39.

Corrigan, P. 1977. Feudal relics or capitalist monuments? Notes on the sociology of unfree labour. *Sociology* **11**, 435–63.

Corrigan, P., H. Ramsay, D. Sayer 1978. *Socialist construction and marxist theory*. London: Macmillan.

Corrigan, P. & D. Sayer 1981. How the law rules: variations on some themes in Karl Marx. In *Law, state and society*, B. Fryer et al. (eds), 21–53. London: Croom Helm.

——1985. *The great arch*. Oxford: Blackwell.

Cumings, B. 1987. The origins and development of the North East Asian political economy: industrial sectors, product cycles and political consequences. See Deyo (1987a), 44–83.

Cutler, A., B. Hindess, P. Hirst, A. Hussain 1977 and 1978. *Marx's capital and capitalism today* [2 volumes]. London: Routledge.

Cypher, J. 1979. The internationalisation of capital and the transformation of social formations: a critique of the Monthly Review School. *Review of Radical Political Economics* **11**(4), 33–49.

Day, R. 1977. Trotsky and Preobrazhensky: the troubled unity of the Left Opposition. *Studies in Comparative Communism* **10**, 77–91.

Delius, P. 1980. Migrant labour and the pedi, 1840–1880. In *Economy and society in pre-industrial South Africa*, S. Marks & A. Atmore (eds), 293–312. London: Longman.

Derrida, J. 1989. Afterword: towards an ethics of discussion. In *Limited Inc.* 111–60. Evanston, Ill.: Northwestern University Press.

Desai, M. 1975. India: emerging contradictions of slow capitalist development. In *Explosion in a subcontinent*, R. Blackburn (ed.), 11–50. London: Penguin.

Dews, P. 1987. *The logics of disintegration*. London: Verso.

Deyo, F. (ed.) 1987a. *The political economy of the new Asian industrialism*. Ithaca: Cornell University Press.

Deyo, F. 1987b. Coalitions, institutions and linkage sequencing – toward a strategic capacity model of East Asian development. See Deyo (1987a), 227–47.

Diamond, I. & L. Quinby (eds) 1988. *Feminism and Foucault*. Boston: Northeastern University Press.

Dore, E. & J. Weeks 1979. International exchange and the causes of backwardness. *Latin American Perspectives* **6**, 62–87.

Doyal, L. & I. Gough 1991. *A theory of human need*. London: Macmillan.

Drakakis-Smith, D. 1992. *Pacific Asia*. London: Routledge.

DuBois, M. 1991. The governance of the third world: a foucauldian perspective on power relations in development. *Alternatives* **16**, 1–30.

Dunning, J. 1981. *International production and the multinational enter-*

prise. London: Allen & Unwin.

Duquette, D. 1992. A critique of the technological interpretation of histori-cal materialism. *Philosophy of the Social Sciences* **22**, 157–86.

Durkheim, E. 1957. *Professional ethics and civil morals*. London: Routledge.

Edwards, C. 1985. *The fragmented world*. London: Methuen.

——1992. Industrialization in South Korea. See Hewitt et al. (1992), 97–127.

Elliott, G. 1992. A just war? The war and the moral gulf. *Radical Philosophy* **61**, 10–13.

Elson, D. 1992. Gender analysis and development economics. Paper pre-sented at ESRC development Economics Study Group Annual Conference, 1–29.

Elson, D. & R. Pearson 1981. Nimble fingers make cheap workers: an analysis of women's employment in third world manufacturing. *Feminist Review* 7, 87–107.

Emmanuel, A. 1972. *Unequal exchange*. London: New Left Books.

Evans, D. 1993. Visible and invisible hands in trade policy reform. See Colclough & Manor (1993), 48–77.

Evans, P. 1987. Class, state and dependence in East Asia: lessons for Latin Americanists. See Deyo (1987a), 203–26.

Evans, P. & J. Stephens 1988. Studying development since the sixties. *Theory and Society* **17**, 713–45.

Fortune 1992. 27 July.

Foster Carter, A. 1978. The modes of production controversy. *New Left Review* **107**, 47–77.

Foucault, M. 1980. *Power/knowledge*. Brighton: Harvester.

Frank, A. G. 1966. The development of underdevelopment. *Monthly Review* **18**, 17–31.

——1969a. *Capitalism and underdevelopment in Latin America*. New York: Monthly Review Press.

——1969b. *Latin America: underdevelopment or revolution?* New York: Monthly Review Press.

——1972. *Lumpenbourgeoisie, lumpendevelopment: dependence, class and politics in Latin America*. New York: Monthly Review Press.

——1981a. *Crisis in the third world*. London: Heinemann.

——1981b. *Reflections on the world economic crisis*. New York: Monthly Review Press.

——1982a. Crisis of ideology and ideology of crisis. See Amin et al. (1982), 128–60.

——1982b. Asia's exclusive models. *Far Eastern Economic Review*, 25 June, 22–3.

——1983. Global crisis and transformation. *Development and Change* **14**,

323–46.

——1991. No escape from the laws of world economics. *Review of African Political Economy* **50**, 21–32.

Freeman, A. 1991. The economic background and consequences of the Gulf War. See Bresheeth & Yuval-Davis (1991), 153–65.

Friedmann, H. 1990. The origins of third world food dependence. See Bernstein et al. (1990), 13–31.

Frobel, F., J. Heinrichs, O. Kreye 1977. The tendency towards a new international division of labour. *Review* **1**, 73–88.

——1980. *The new international division of labour*. Cambridge: Cambridge University Press.

Fukuyama, F. 1989. The end of history. *The National Interest* **16**, 3–17.

——1992. *The end of history and the last man*. London: Penguin.

Gallie, D. 1983. *Social inequality and class radicalism in France and Britain*. Cambridge: Cambridge University Press.

Gamble, A. 1983. Monetarism and the social market economy. In *Socialist arguments*, D. Coates & G. Johnston (eds), 7–32. London: Martin Robertson.

Geras, N. 1985. The controversy about Marx and justice. *New Left Review* **150**, 47–85.

Gerschenkron, A. 1962. *Economic backwardness in historical perspective*. Cambridge: Belknap Press.

Giddens, A. 1984. *The constitution of society*. Oxford: Blackwell.

Gills, B., J. Rocamora, R. Wilson (eds) 1993. *Low intensity democracy*. London: Pluto.

Goldsmith, E. 1992. *The way: an ecological world view*. London: Rider.

Gordon, D. 1988. The global economy: new edifice or crumbling foundations? *New Left Review* **168**, 24–64.

Gough, J. 1986. Industrial policy and socialist strategy. *Capital and Class* **29**, 58–79.

Gowan, P. 1991. The Gulf War, Iraq and western liberalism. *New Left Review* **187**, 29–70.

Gulalp, H. 1986. Debate on capitalism and development: the theories of Samir Amin and Bill Warren. *Capital and Class* **28**, 135–59.

——1987. Capital accumulation, classes and the relative autonomy of the state. *Science and Society* **51**, 287–311.

Habermas, J. 1987. *The philosophical discourse of modernity*. Cambridge: Polity.

——1992. *Autonomy and solidarity*. London: Verso.

Hall, S. 1992. The west and the rest: discourse and power. In *Formations of modernity*, S. Hall & B. Gieben (eds), 276–320. Cambridge: Polity.

Halliday, F. 1983. *The making of the second cold war*. London: Verso.

——1989. *Cold war, third world*. London: Hutchinson.

——1990. The crisis of the Arab World: the false answers of Saddam Hussein. *New Left Review* **184**, 69–74.

——1991. The left and the war. *New Statesman*, 8 March, 15–16.

Hamilton, C. (n.d.) Capitalist industrialisation in East Asia's four little tigers. In *Neo-Marxist theories of development*, P. Limqueco & B. McFarlane (eds), 137–80. London: Croom Helm.

Hamilton, C. 1986. *Capitalist industrialization in Korea*. London: Westview.

——1987. Can the rest of Asia emulate the NICs? *Third World Quarterly* 87, 1225–56.

Hamilton, P. 1992. The enlightenment and the birth of social science. In *Formations of modernity*, S. Hall & B. Gieben (eds), 18–58. Cambridge: Polity.

Harman, C. 1993. Where is capitalism going? *International Socialism* **60**, 77–136.

Harris, N. 1986. *The end of the Third World*. Harmondsworth: Penguin.

Harriss, B. & B. Crow 1992. Twentieth century free trade reform: food market deregulation in Sub-Saharan Africa and South Asia. In *Development policy and public action*, M. Wuyts, M. Mackintosh, T. Hewitt (eds), 199–227. Oxford: Oxford University Press.

Harriss, J. 1987. Capitalism and peasant production: the green revolution in India. In *Peasants and peasant societies*, T. Shanin (ed.), 227–36. Oxford: Blackwell.

Hart-Landsberg, M. 1979. Export-led industrialization in the third world: manufacturing imperialism. *Review of Radical Political Economics* **16**, 181–93.

——1984. Capitalism and third world economic development: a critical look at the South Korean miracle. *Review of Radical Political Economics* **16**, 181–93.

Harvey, D. 1989. *The condition of postmodernity*. London: Blackwell.

——1993. The nature of environment: dialectics of social and environmental change. In *The Socialist Register 1993*, 1–51. London: Merlin Press.

Hayek, F. 1960. *The constitution of liberty*. London: Routledge & Kegan Paul.

Healey, J. & M. Robinson 1992. *Democracy, governance and economic policy*. London: Overseas Development Institute.

Henderson, J. & R. Appelbaum 1992. Situating the state in the East Asian development process. In *States and development in the Asian Pacific Rim*, R. Appelbaum & J. Henderson (eds), 1–26. Newbury Park: Sage.

Herrnstein-Smith, B. 1988. *Contingencies of value*. Cambridge, Mass.: Harvard University Press.

Hewitt, M. 1993. Illusions of freedom: the regressive implications of postmodernism. In *The Socialist Register 1993*, 78–91. London: Merlin.

Hewitt, T., H. Johnson, D. Wield (eds) 1992. *Industrialization and development*. Oxford: Oxford University Press.

Hilton, R. (ed.) 1976a. *The transition from feudalism to capitalism*. London: Verso.

Hilton, R. 1976b. A comment. See Hilton (1976a), 109–17.

Hindess, B. 1978. Classes and politics in Marxist theory. In *Power and the state*, G. Littlejohn (ed.), 72–97. London: Croom Helm.

Hindess, B. & P. Hirst 1975. *Precapitalist modes of production*. London: Routledge.

——1977. *Mode of production and social transformation*. London: Macmillan.

Hirst, P. 1977. Economic classes and politics. In *Class and class structure*, A. Hunt (ed.), 125–54. London: Lawrence & Wishart.

Hitchens, C. 1993. *For the sake of argument*. London: Verso.

Hobsbawm, E. 1968. *Industry and empire*. London: Weidenfeld & Nicholson.

Holloway, J. & S. Picciotto (eds) 1978. *State and capital: a Marxist debate*. London: Edward Arnold.

Hoogvelt, A. 1982. *The third world in global development*. London: Macmillan.

Hoselitz, B. (ed.) 1960. *The sociological aspects of economic growth*. New York: Free Press.

Howe, S. 1992. Empire strikes back. *New Statesman* 9 (October), 36–7.

Hughes, R. 1993. *The culture of complaint*. Oxford: Oxford University Press.

Hulme, P. 1986. *Colonial encounters: Europe and the native Caribbean, 1492–1797*. London: Methuen.

Hyden, G. 1983. *No shortcuts to progress*. London: Heinemann.

Institute of Development Studies 1987. Cyprus industrial strategy: main report. *Report of UNDF/UNIDO Mission*, Mimeo.

International Monetary Fund 1989. *World economic outlook*. Washington, DC: International Monetary Fund.

de Janvry, A. 1987. Peasants, capitalism and the state in Latin America culture. In *Peasants and Peasant Societies*, T. Shanin (ed.), 390–8. Oxford: Blackwell.

Jenkins, R. 1984a. Divisions over the international division of labour. *Capital and Class* **22**, 28–57.

——1984b. *Transnational corporations and the industrial transformation of Latin America*. London: Macmillan.

——1987. *Transnational corporations and uneven development*. London: Methuen.

——1990. Learning from the gang: are there lessons for Latin America from East Asia. *Bulletin of Latin American Research* **10**, 37–54.

——1991. The political economy of industrialization: a comparison of Latin American and East Asian newly industrializing countries. *Development and Change* **22**, 197–231.

——1992. (Re-)interpreting Brazil and South Korea. See Hewitt, Johnson, Wield (1992), 167–98.

Kaplinsky, R. 1993. Industrialization in Botswana: how getting the prices right helped the wrong people. See Colclough & Manor (1993), 148–72.

Kerr, C., J. Dunlop, F. Harbison, C. Myers 1962. *Industrialism and industrial man*. London: Heinemann.

Kiely, R. 1992a. Marxism and the sociology of development: explanations for the impasse. Warwick Working Papers in Sociology no.18, Coventry: University of Warwick, Department of Sociology.

——1992b. The "third world" and the "new world order". Paper to Conference of Socialist Economists.

——1994. Development theory and industrialisation: beyond the impasse. *Journal of Contemporary Asia* **24**, 133–60.

——1995. Third Worldist relativism: a new form of imperialism. *Journal of Contemporary Asia* **25**, 00–00.

Kiernan, V. 1980. *America: the new imperialism*. London: Zed.

Kitching, G. 1989. *Development and underdevelopment in historical perspective*. London: Routledge.

Kowalik, T. 1991. Marketization and privatization: the Polish case. In *The Socialist Register 1991*, 259–78. London: Merlin.

Kreuger, A. 1974. The political economy of the rent seeking society. *American Economic Review* **64**, 291–303.

Laclau, E. 1971/1977. Feudalism and capitalism in Latin America. *New Left Review* **67**, 19–38; and in *Politics and ideology in Marxist theory*, 15–50. London: Verso.

——1990. *New reflections on the revolution in our time*. London: Verso.

Laclau, E. & C. Mouffe 1985. *Hegemony and socialist strategy*. London: Verso.

Lal, D. 1983. *The poverty of "development economics"*. London: Institute of Economic Affairs.

Larrain, J. 1986. *A reconstruction of historical materialism*. London: Allen & Unwin.

——1989. *Theories of development*. Cambridge: Polity.

——1994. *Ideology and cultural identity*. Cambridge: Polity.

Laurell, A. C. 1992. Democracy in Mexico: will the first be the last. *New Left Review* **194**, 33–53.

Leftwich, A. 1993. Governance, democracy and development in the third world. *Third World Quarterly* **14**, 605–24.

Lenin, V. I. 1977. *Selected works*. Moscow: Progress.

Levine, A. & E. O. Wright 1980. Rationality and class struggle. *New Left*

Review **123**, 47–68.

Lewis, W. A. 1950. *The industrialisation of the British West Indies.* West Indies.

Leys, C. 1982. African economic development in theory and practice. *Daedalus* **111**, 99–124.

——1986. Conflict and convergence in development theory. In *Imperialism and after*, W. Mommsen & J. Osterhammel (eds), 315–24. London: Allen & Unwin.

Liebman, M. 1980. *Leninism under Lenin.* London: Merlin.

Linebaugh, P. 1993. *The London hanged.* Harmondsworth: Penguin.

Lipietz, A. 1986. New tendencies in the international division of labour: regimes of accumulation and modes of regulation. In *Production, work, territory*, A. Scott & M. Storper (eds), 16–40. London: Allen & Unwin.

——1987. *Miracles and mirages.* London: Verso.

Little, I. 1981. *Economic development.* New York: Basic.

Littler, C. 1984. Soviet type societies and the labour process. In *Work, employment and unemployment*, K. Thompson (ed.), 87–96. Milton Keynes: Open University.

Long, N. & M. Villareal 1993. Exploring development interfaces: from the transfer of knowledge to the transfer of meaning. See Schuurman (1993), 187–206.

Lovibond, S. 1992. Feminism and pragmatism: a reply to Richard Rorty. *New Left Review* **193**, 56–74.

Lowy, M. 1977. Marxism and the national question. In *Revolution and class struggle*, R. Blackburn (ed.), 136–63. London: Fontana.

——1981. *The politics of combined and uneven development.* London: Verso.

Lyotard, J-F. 1984. *The post-modern condition.* Manchester: Manchester University Press.

McClelland, D. 1961. *The achieving society.* New York: Free Press.

McGrew, A. 1992. The third world in the new global order. In *Poverty and development in the 1990s*, T. Allen & A. Thomas (eds), 255–72. Oxford: Oxford University Press.

Mackintosh, M. 1977. Reproduction and patriarchy: a critique of Meillassoux' "femmes, greniers et capitaux". *Capital and Class* **2**, 119–27.

Mackintosh, M. 1990. Abstract markets and real needs. See Bernstein et al. (1990), 43–53.

McLennan, G. 1992. The enlightenment project revisited. In *Modernity and its futures*, S. Hall, D. Held, T. McGrew (eds), 327–55. Cambridge: Polity.

Magas, B. 1993. Once more on Yugoslavia – a reply to Iraj Hashi and Michael Barratt-Brown. *Capital and Class* **50**, 161–7.

Mandel, D. 1989. "Revolutionary reform" in Soviet factories. In *The Socialist Register 1989*, 102–209. London: Merlin.

Mandle, J. 1972. The plantation economy: an essay in definition. *Science and Society* **36**, 49–62.

Mann, M. 1987. The social cohesion of liberal democracy. In *Society and the social sciences*, D. Potter (ed.), 255–68. London: Routledge.

Marglin, S. (ed.) 1991. *Dominating knowledge: development, culture and resistance.* Oxford: Clarendon.

Marx, K. 1967. *Capital*, vol. 2. Moscow: Progress Publishers.

——1973. *Grundrisse.* London: Penguin.

——1976a. *The poverty of philosophy.* Peking: Foreign Languages Press.

——1976b. *Capital*, vol. 1. Harmondsworth: Penguin.

——1977. *Selected writings.* Oxford: Oxford University Press.

——1982. Pathways of social development: a brief against suprahistorical theory. See Alavi & Shanin (1982), 109–11.

——1984. The reply to Zasulich. See Shanin (1984), 123–6.

——1989. *Readings.* London: Routledge (ed. D. Sayer).

Marx, K. & F. Engels 1974. *On colonialism.* Moscow: Progress.

——1982. *The German ideology.* London: Lawrence & Wishart.

Meiksins Wood, E. 1981. The separation of the economic and political in capitalism. *New Left Review* **127**, 66–95.

——1984. Marxism and the course of history. *New Left Review* **147**, 95–107.

——1986. *The retreat from class.* London: Verso.

——1990. Falling through the cracks: E. P. Thompson and the debate on base and superstructure. In *E. P. Thompson: critical perspectives*, H. Kaye & K. McClelland (eds), 125–52. Cambridge: Polity.

——1991. *The pristine culture of capitalism.* London: Verso.

Meillassoux, C. 1981. *Maidens, meals and money.* Cambridge: Cambridge University Press.

Miles, R. 1987. *Capitalism and unfree labour.* London: Tavistock.

Mitter, S. 1986. *Common fate, common bond.* London: Pluto.

Moore, M. 1993a. Introduction. *IDS Bulletin* **24**, 1–6.

——1993b. Declining to learn from the east? The World Bank on governance and development. *IDS Bulletin* **24**, 39–50.

Moore, W. 1963. *Social change.* New Jersey: Prentice-Hall.

——1965. *The impact of industry.* New Jersey: Prentice-Hall.

Moser, C. 1978. Informal sector or petty commodity production: dualism or dependency in urban development. *World Development* **6**, 1041–64.

Mosley, P., J. Harrigan, J. Toye 1991. *Aid and power* [2 volumes]. London: Routledge.

Moss, R. 1975. *The collapse of democracy.* London: Institute of Economic Affairs.

Mouzelis, N. 1980. Modernization, underdevelopment, uneven development: prospects for a theory of third world formations. *Journal of Peasant*

Studies 7, 353–74.

——1988. Sociology of development: reflections on the present crisis. *Sociology* 22, 23–44.

——1990. *Post-Marxist alternatives*. London: Macmillan.

Mueller, S. 1980. Retarded capitalism in Tanzania. In *The Socialist Register 1980*, 203–26. London: Merlin.

Munslow, B. 1983. Why has the Westminster model failed in Africa? *Parliamentary Affairs* 36, 164–80.

——1993. Democratisation in Africa. *Parliamentary Affairs* 46, 478–90.

Murray, F. 1987. Flexible specialisation in the "third Italy". *Capital and Class* 33, 84–95.

Murray, R. (ed.) 1981. *Multinationals beyond the market*. Brighton: Harvester.

Naishul, V. 1990. Problems of creating a market in the USSR. *Communist Economies* 2, 275–90.

New Internationalist 1992. The New Globalism, **246**.

New Statesman 1991. Leader, 22 March.

Norris, C. 1990. *What's wrong with postmodernism?* London: Harvester-Wheatsheaf.

——1992. *Uncritical theory*. London: Lawrence & Wishart.

——1993. Old themes for new times: Basildon revisited. In *The Socialist Register 1993*, 52–77. London: Merlin.

Nussbaum, M. 1992. Human functioning and social justice. *Political Theory* 20, 202–46.

O'Brien, P. 1975. A critique of Latin American theories of dependency. In *Beyond the sociology of development*, I. Oxaal, T. Barnett, D. Booth (eds), 7–27. London: Routledge & Kegan Paul.

O'Donnell, K. & P. Nolan 1989. Flexible specialisation and the Cyprus industrial strategy. *Cyprus Journal of Economics* 2, 1–20.

Ogle, G. 1990. *South Korea: dissent within the economic miracle*. London: Zed.

Ohlin, B. 1933. *Inter-regional and international trade*. Cambridge, Mass.: Harvard University Press.

Ollman, B. 1976. *Alienation*. Cambridge: Cambridge University Press.

Oxaal, I., T. Barnett, D. Booth (eds) 1975. *Beyond the sociology of development*. London: Routledge & Kegan Paul.

Palloix, C. 1975. The internationalization of capital and the circuit of social capital. In *International firms and modern imperialism*, H. Radice (ed.), 63–88. Harmondsworth: Penguin.

Palma, G. 1978. Dependency and development: a formal theory of underdevelopment or a methodology for the analysis of concrete situation of underdevelopment. *World Development* 6, 881–924.

Parpart, J. 1993. Who is the other? A postmodern feminist critique of women and development theory and practice. *Development and Change* **24**, 439–64.

Parsons, T., R. Bales, E. Shils 1962. *Working papers on the theory of action.* London: Collier-Macmillan.

Pearce, J. 1982. *Under the eagle.* London: Latin America Bureau.

Peet, R. 1986. Industrial devolution and the crisis of international capitalism. *Antipode* **18**, 78–95.

——1989. Conceptual problems in neo-Marxist industrial geography: a critique of themes from Scott and Storper's "Production, work and territory". *Antipode* **21**, 35–50.

——1991. *Global capitalism.* London: Routledge.

——1992. Some critical questions for anti-essentialism. *Antipode* **24**, 113–30.

Petras, J. & H. Brill 1985. The tyranny of globalism. *Journal of Contemporary Asia* **15**, 20–32.

Phillips, A. 1977. The concept of development. *Review of African Political Economy* **8**, 7–20.

Pilger, J. 1993. Intervention as imperialism. *New Statesman*, 22 January.

Pilling, G. 1973. Imperialism, trade and unequal exchange: the work of Arghiri Emmanuel. *Economy and Society* **2**, 164–86.

Pinkney, R. 1993. *Democracy in the third world.* Buckingham: Open University Press.

Piore, M. & C. Sabel 1984. *The second industrial divide.* New York: Basic.

Plekhanov, G. 1976. *The materialist conception of history.* London: Lawrence & Wishart.

Popper, K. 1986. *The poverty of historicism.* London: Ark.

Post, K. 1978. *Arise ye starvelings! The Jamaican labour rebellion of 1938 and its aftermath.* The Hague: Martinus Nijhoff.

Poulantzas, N. 1973. *Political power and social classes.* London: Verso.

Prebisch, R. 1959. Commercial policy in the underdeveloped countries. *American Economic Review* **44**, 251–73.

Raikes, P. 1988. *Modernising hunger.* London: Catholic Institute of International Relations.

Rajamoorthy, T. 1993. World human rights conference – a view from the south. *Third World Resurgence* **36**.

Rasmussen, J., H. Schmitz, P. van Dijk 1992. Introduction: exploring a new approach to small scale industry. *IDS Bulletin* **23**, 2–7.

Ricardo, D. 1971. *Principles of political economy.* Cambridge: Cambridge University Press.

Roddick, J. 1988. *The dance of the millions.* London: Latin America Bureau.

Rodney, Walter 1972. *How Europe underdeveloped Africa.* London: Bogle L'Ouverture.

Rorty, R. 1991. Feminism and pragmatism. *Michigan Quarterly Review* **30**,

231–58.

Rostow, W. 1960/1971. *The stages of economic growth*. Cambridge: Cambridge University Press.

Ruccio, D. & L. Simon 1986. Methodological aspects of a Marxian approach to development: an analysis of the modes of production school. *World Development* **14**, 211–22.

Runciman, W. G. 1983. *A treatise on social theory*, vol. 1. Cambridge: Cambridge University Press.

Sabel, C. 1982. *Work and politics*. Cambridge: Cambridge University Press.

——1986. Changing models of economic efficiency and their implications for industrialisation in the third world. In *Development, democracy and the art of trespassing*, A. Foxley, M. McPherson, G. O'Donnell (eds). Notre Dame, Ind.: University of Notre Dame Press.

Sachs, J. 1984. Comments on Diaz-Alejandro. *Brookings Papers on Economic Activity*, **2**, 393–401.

Sachs, W. 1992. Development: a guide to the ruins. *New Internationalist* **232**.

Said, E. 1978. *Orientalism*. London: Penguin.

——1993a. *Culture and imperialism*. London: Chatto & Windus.

——1993b. Orientalism and after: an interview with Edward Said. *Radical Philosophy* **63**, 22–32.

Samara, A. 1991. The new international order. See Bresheeth & Yuval-Davis (1991), 259–71.

Sandbrook, R. 1986. *The politics of Africa's economic stagnation*. Cambridge: Cambridge University Press.

——1993. *The politics of Africa's economic recovery*. Cambridge: Cambridge University Press.

Saul, J. 1993. Rethinking the Frelimo State. In *The Socialist Register 1993*, 139–65. London: Merlin.

Sayer, D. 1975. Method and dogma in historical materialism. *Sociological Review* **23**, 784–811.

——1979. Science as critique: Marx versus Althusser. In *Issues in Marxist philosophy*, vol. 3, J. Mepham & D. Ruben (eds), 85–110. Brighton: Harvester.

——1987. *The violence of abstraction*. Oxford: Blackwell.

——1991. *Capitalism and modernity*. London: Routledge.

——1992. A notable administration: English state formation and the rise of capitalism. *American Journal of Sociology* **97**, 1382–1416.

Schiffer, J. 1981. The changing post-war pattern of development. *World Development* **9**, 715–28.

Schuurman, F. (ed.) 1993. *Beyond the impasse: new directions in development theory*. London: Zed.

Scott, A. 1988. *New industrial spaces*. London: Pion.

Seers, D. 1979. The congruence of Marxism and other neoclassical doctrines. In *Toward a new strategy for development*, A. Hurschmann (ed.), 1–17. Oxford: Pergamon.

Selden, M. & Chih-ming Ka 1988. Original accumulation, equality and late industrialization: the cases of socialist China and capitalist Taiwan. In *The political economy of chinese socialism*, M. Selden, 101–28. London: M. E. Sharpe.

Sen, A. 1981. *Poverty and famines* Oxford: Oxford University Press.

——1984. *Resources, values and development*. Oxford: Blackwell.

Sender, J. & S. Smith 1985. What's right with the Berg Report and what's left of its critics. *Capital and Class* **24**, 125–46.

——1986. *The development of capitalism in Africa*. London, Methuen.

——1990. *Poverty, class and gender in rural Africa*. London: Routledge.

Shaikh, A. 1980. Foreign trade and the law of value: part two. *Science and Society* **44**, 27–57.

Shanin, T. (ed.) 1984. *Late Marx and the Russian road*. London: Routledge.

Shaw, M. 1993. Grasping the nettle. *New Statesman*, 15 January.

Sklair, L. 1988. Transcending the impasse: metatheory, theory and empirical research in the sociology of development and underdevelopment. *World Development* **16**, 697–709.

Slater, D. 1992. Theories of development and politics of the post-modern – exploring a border zone. *Development and Change* **23**, 283–319.

Smith, A. 1910. *The wealth of nations*. London: Everyman.

Smith, R. 1993. The Chinese Road to capitalism. *New Left Review* **199**, 55–99.

Soper, K. 1993. A theory of human need. *New Left Review* **197**, 113–28.

Southall, R. 1988. Introduction. In *Trade unions and the new industrialisation of the third world*, R. Southall (ed.). London: Zed.

Spero, J. 1990. *The politics of international economic relations*, 4th edn. London: Unwin Hyman.

Spraos, J. 1983. *Inequalising trade?* Oxford: Clarendon.

Spybey, T. 1992. *Social change, development and dependency*. Cambridge: Polity.

Squires, J. (ed.) 1993. *Principled positions*. London: Lawrence & Wishart.

Stalin, J. V. 1976. *Problems of Leninism*. Peking: Foreign Languages Press.

Stork, J. & A. Lesch 1990. Why war? *Middle East Report* **167**, 11–18.

Taylor, J. 1979. *From modernisation to modes of production*. London: Macmillan.

Tenbruck, F. 1990. The dream of a secular ecumene: the meaning and limits of policies of development. In *Global culture*, M. Featherstone (ed.), 193–206. London: Sage.

Thatcher, M. 1986. What has gone wrong? In *The economic decline of modern Britain*, D. Coates & J. Hillard (eds), 64–6. London: Wheatsheaf.

Thomas, A. 1983. Third world: images, definitions, connotations. In Open University U204, *Third World Studies*, Block 1, 1–45.

Thomas, C. 1977. The non-capitalist path as theory and practice of decolonization and socialist transformation. *Latin American Perspectives* 17, 10–28.

Thompson, E. P. 1963. *The making of the English working class*. London: Gollancz.

——1965. The peculiarities of the English. In *The poverty of theory*, E. P. Thompson, 245–301. London: Merlin, 1975.

Toye, J. 1985. Dirigisme and development economics. *Cambridge Journal of Economics* 9, 1–14.

——1987. *Dilemmas of development*. Oxford: Blackwell.

United Nations Centre on Transnational Corporations 1992. *World investment report*. New York: UNCTC.

Urry, J. 1989. The end of organised capitalism. In *New times*, M. Jacques & S. Hall (eds), 94–102. London: Lawrence & Wishart.

van der Geest, P. & F. Buttel 1988. Marx, Weber and development. Sociology: beyond the impasse. *World Development* 16, 683–95.

Wada, H. 1984. Marx and revolutionary Russia. See Shanin (1984), 40–75.

Wade, R. 1983. South Korea's agricultural development: the myth of the passive state. *Pacific Viewpoint* 24, 11–29.

Wade, R. 1990. *Governing the market*. Princeton, NJ: Princeton University Press.

Wallerstein, I. 1974. *The modern world system*, vol. 1. New York: Academic Press.

Wallerstein, I. (ed.) 1983. *Labour in the world social structure*. Beverly Hills, Calif.: Sage.

Wallerstein, I. 1986. Braudel on capitalism and the market. *Monthly Review* 37, 10–21.

Walzer, M. 1977. *Just and unjust wars*. London: Allen Lane.

——1992. *Just and unjust wars*, 2nd edn. London: Basic.

Warren, B. 1973. Imperialism and capitalist industrialisation. *New Left Review* 81, 9–44.

——1980. *Imperialism: pioneer of capitalism*. London: Verso.

Watts, M. 1990. Peasants under contract: agro-food complexes in the third world. See Bernstein et al. (1990), 69–79.

Webster, A. 1990. *Introduction to the sociology of development*. London: Macmillan.

Weeks, J. 1982. Equilibrium, uneven development and the tendency of the rate of profit to fall. *Capital and Class* 16, 62–77.

White, G. 1992. Changing patterns of public action in socialist development: the Chinese decollectivization. In *Development policy and public action*, M. Wuyts, M. Mackintosh and T. Hewitt (eds), 231–52. Oxford:

Oxford University Press.

Willetts, P. 1978. *The non-aligned movement*. London: Pinter.

Williams, E. 1987. *Capitalism and slavery*. London: Andre Deutsch.

Williams, G. 1978a. Imperialism and development: a critique. *World Development* **6**, 925–36.

——1978b. In defence of history. *History Workshop Journal* **7**, 116–24.

——1987. Primitive accumulation: the way to progress? *Development and Change* **18**, 637–59.

——1988. Why is there no agrarian capitalism in Nigeria. *Journal of Historical Sociology* **1**, 345–98.

Williams, K., T. Cutler, J. Williams, C. Haslam 1987. Review article: the end of mass production. *Economy and Society* **16**, 405–38.

Wolpe, H. 1972. Capitalism and cheap labour-power in South Africa: from segregation to apartheid. *Economy and Society* **1**, 425–56.

Wolpe, H. (ed.) 1980. *The articulation of modes of production*. London: Routledge.

Wolpe, H. 1988. *Race, class and the apartheid state*. London: James Currey.

World Bank 1981. *Accelerated development in Sub-Saharan Africa: an agenda for action*. Washington, DC: World Bank.

——1983. *World development report 1983*. Oxford: Oxford University Press.

——1984. *Towards sustained development in Sub-Saharan Africa*. Washington, DC: World Bank.

——1985. *World development report 1985*. Oxford: Oxford University Press.

——1988. *World development report 1988*. Oxford: Oxford University Press.

——1989. *Sub-Saharan Africa: from crisis to sustainable growth*. Washington, DC: World Bank.

——1992. *Governance and development*. Washington, DC: World Bank.

Worsley, P. 1964. *The third world*. London: Weidenfeld & Nicholson.

——1984. *The three worlds*. London: Weidenfeld & Nicholson.

Young, K. 1993. *Planning development with women*. London: Macmillan.

Index